POLYMER SCIENCE AND TECHNOLOGY

T0076003

ASYMMETRIC HETEROGENEOUS SUPPORTED CATALYSIS

USE OF NITROGEN-CONTAINING LIGANDS

POLYMER SCIENCE AND TECHNOLOGY

Additional books in this series can be found on Nova's website under the Series tab.

Additional E-books in this series can be found on Nova's website under the E-book tab.

POLYMER SCIENCE AND TECHNOLOGY

ASYMMETRIC HETEROGENEOUS SUPPORTED CATALYSIS

USE OF NITROGEN-CONTAINING LIGANDS

CHRISTINE SALUZZO

AND

STÉPHANE GUILLARME

Nova Science Publishers, Inc.

New York

LIBRARY OF CONGRESS CATALOGING-IN-PUBLICATION DATA

Saluzzo, Christine.
  Asymmetric heterogeneous supported catalysis : use of nitrogen-containing ligands / Christine Saluzzo and Stiphane Guillarme.
    p. cm.
  Includes index.
  ISBN 978-1-61668-680-2 (softcover)
  1. Enantioselective catalysis. 2. Asymmetric synthesis. 3. Ligands. 4. Nitrogen compounds. I. Guillarme, Stiphane. II. Title.
  QD505.S2545 2009
  541'.395--dc22
                          2010016689

Published by Nova Science Publishers, Inc. † New York

CONTENTS

Preface		**vii**
Chapter 1	Introduction	**1**
Chapter 2	Reduction of C=O and C=N Bond	**3**
Chapter 3	Cyclopropanation	**35**
Chapter 4	Cycloadditions and Heterocycloadditions	**55**
Chapter 5	Addition of Organometallic Reagents to Aldehydes, Ketones and Imines	**65**
Chapter 6	Asymmetric π-Allylic Substitution	**109**
Chapter 7	Dihydroxylation and Aminohydroxylation	**113**
Chapter 8	Epoxidation	**129**
Chapter 9	Kinetic Resolution of Terminal Epoxides	**145**
Chapter 10	Miscellaneous	**153**
Chapter 11	Conclusion	**157**
References		**159**
Index		**171**

PREFACE

The need to develop effective methods for enantioselective synthesis is becoming ever more important as only a single enantiomer of a racemic bioactive compound is generally required for pharmaceuticals, agrochemicals, flavour or fragrance. Homogeneous asymmetric catalysis has been used to perform a variety of transformations under mild conditions with high enantioselectivity. As these fine and specialty chemicals are manufactured to meet high and well defined standards of purity compatible with the desired performance, chemical industry had to adopt efficient and clean technology. This book discusses the contribution of soluble polymer supported ligands and insoluble polymer-supported ligands to asymmetric catalysis in the field of reduction of C=O and C=N bonds, cyclopropanation, Diels-Alder, alkylation, allylation, dihydroxylations, epoxydation reactions, kinetic resolution of terminal epoxide by means of nitrogen containing ligands complexed with metal as asymmetric catalyst. Each type of asymmetric homogeneous supported catalyst will be examined successively from the perspective of using phosphine-free nitrogen- containing ligands complexed with a metal. Results were analyzed in comparison to the results in the literature for similar systems.

Chapter 1

INTRODUCTION

The need to develop effective methods for enantioselective synthesis is becoming ever more important as only a single enantiomer of a racemic bioactive compound is generally required for pharmaceuticals [1], agrochemicals [1,2], flavour [1,3] or fragrance [1,3]. Homogeneous asymmetric catalysis has been used to perform a variety of transformations under mild conditions with high enantioselectivity. As these fine and specialty chemicals are manufactured to meet high and well defined standards of purity compatible with the desired performance, chemical industry had to adopt efficient and clean technology [4]. From an industrial point of view, homogeneous methods remain unpratical, particularly due to the high cost of the chiral metal catalysts (range of the chiral catalyst loadings from 1 to 10% and in some cases up to 30%) and the difficulty of their recovery and reuse. Thus, from an economic, environmental and technical point of view, homogeneous supported catalysis is preferable to homogeneous catalysis because of the handling, separation and recycling abilities [5]. Many approaches have been employed to immobilize the homogeneous catalyst. [6] In the classical immobilization with organic polymer, the chiral ligand units are anchored onto polymers (type 1, Scheme 1). The incorporation of chiral ligands on the main chain of the polymers is another approach (type 2, Scheme 1) and is done by copolymerization of the chiral ligand with a linker. The last but less known approach consists in the use of an imprinted polymer (type 3, Scheme 1), an extensively used methodology to form artificial antibodies. The imprinted polymer is based on a copolymerization and a crosslinking of a polymer around a molecule used as a printed molecule (PM). This latter included into the polymer by means of an interaction with functional groups is then removed, thus leaving its imprint in the polymeric matrix. If a chiral PM is used, the formed chiral cavity is able to act as a center of molecular recognition [7].

The aim of this review is to discuss the contribution of soluble polymer-supported ligands and insoluble polymer-supported ligands to asymmetric catalysis in the field of reduction of C = O bonds, cyclopropanation, Diels-Alder, alkylation, allylation, dihydroxylations, epoxydation reactions, kinetic resolution of terminal epoxide ..... by means of nitrogen containing ligands complexed with metal as asymmetric catalyst.

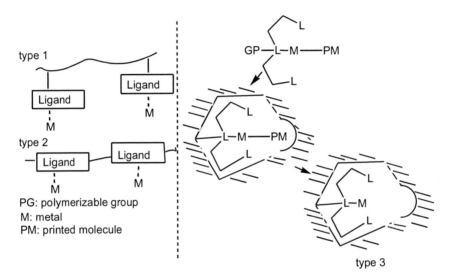

Scheme 1.

Each type of asymmetric homogeneous supported catalyst will be examined successively from the perspective of using phosphine-free nitrogen-containing ligands complexed with a metal.

Results were analyzed in comparison to the results in the literature for similar systems.

Chapter 2

REDUCTION OF C=O AND C=N BOND

The three major catalytic reduction procedures which have emerged are enantioselective hydride reduction, hydrogenation and hydrogenation transfer reduction (HTR) close to the hydrosilylation which represents only few examples in asymmetric polymer-supported catalyst.

2.1. HYDROGENATION

In homogeneous catalysis, various phosphines have been developped for asymmetric hydrogenation [8]. However, for hydrogenation of unfunctionalized ketone, a remarkably high reactivity emerges when the Ru compounds are further complexed with a 1,2 diamine ligand. Thus, the precatalyst $RuCl_2(phosphine)_2(1,2$-diamine) combined with an inorganic base in isopropanol is considered as one of the most powerful systems for the hydrogenation of ketones (Scheme 2, *i*-PrOH as solvent) [9, 10].

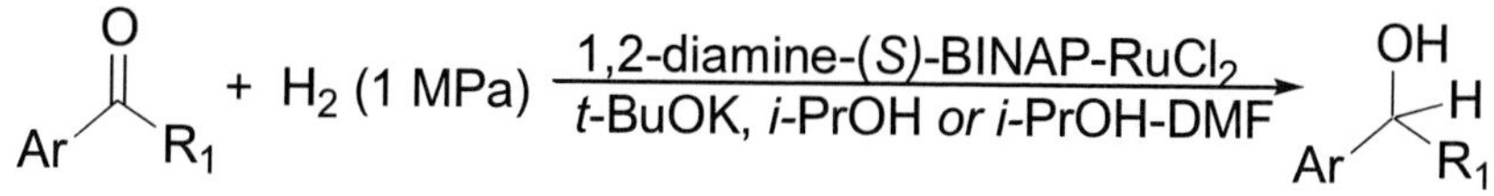

$$\underset{\text{Ar}}{\overset{\text{O}}{\|}}\underset{\text{R}_1}{\diagdown} + \text{H}_2 \text{ (1 MPa)} \xrightarrow[\text{\textit{t}-BuOK, \textit{i}-PrOH or \textit{i}-PrOH-DMF}]{\text{1,2-diamine-(\textit{S})-BINAP-RuCl}_2} \underset{\text{Ar}}{\overset{\text{OH}}{\diagup}}\underset{\text{R}_1}{\overset{\text{H}}{\diagdown}}$$

Scheme 2.

Close to polymer bound-BINAP/diamine ruthenium precatalyst [11], complexes of BINAP or their derivatives in the presence of polymer bound diamine have been employed. These catalytic systems have been mainly studied by Itsuno with immobilized diphenylethylenediamine (DPEN). As the use of *N*-substituted 1,2-diamine ligands decreased the catalytic activity of hydrogenation of simple ketones contrary to free diamine ligands, they have been less employed. However, *N*-substituted 1,2-diamine ligands are rather effective in the hydrogenation of α-substituted ketone, such as α-amido ketones. Thus, two types of immobilized DPEN have been performed. Functionalized DPEN for subsequent grafting onto polymer or copolymerization are presented (Scheme 3).

Homogeneous ligands   Functionalized DPEN for subsequent grafting or copolymerization

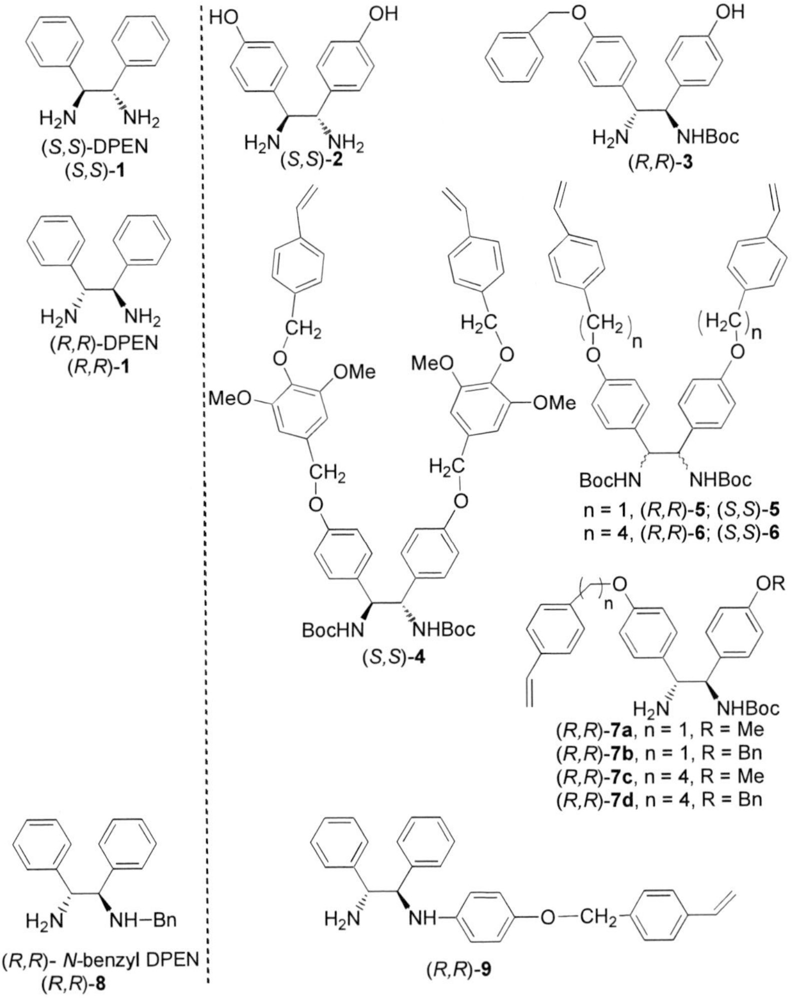

Scheme 3.

If free primary amines were maintained, grafting was realized by the modification of phenyl groups into phenolic ones [12, 13] and copolymerization by using a phenolic ether group containing a vinylphenyl functionality as monomer. ((S,S)-2, (R,R)-3 and (S,S)-4, -5, -6; (R,R)-5, -6, -7, -9 respectively (Scheme 3) [13, 14, 15].

The first type of polymers was formed by reaction of the phenoxide derived from DPEN with 1% crosslinked chloromethylated polystyrene (Schemes 4 and 5). It has been observed that in the cases of higher content of chloromethyl group or a higher degree of crosslinking, some unreacted chloromethyl groups remain.

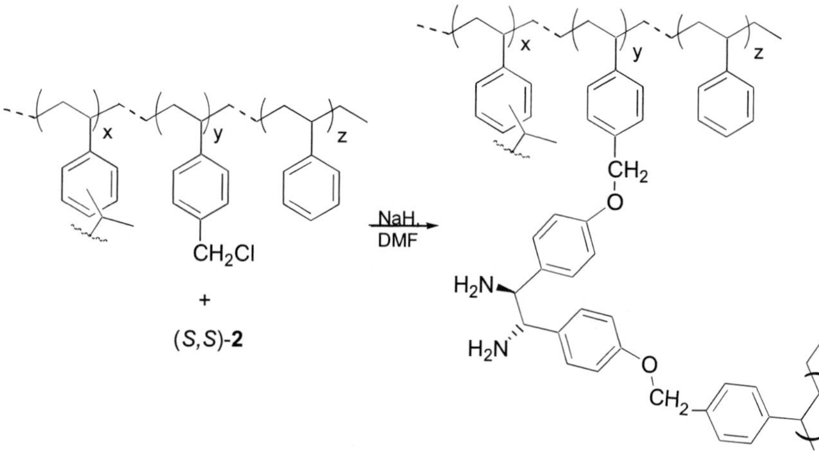

(S,S)-10a: x = 0.01, y = 0.2, z = 0.79
(S,S)-10b: x = 0.05, y = 0.2, z = 0.75

Scheme 4.

To perform the catalytic homogeneous-supported hydrogenation of acetophenone with precatalyst ((S,S)-10a, (S,S)-10b, (R,R)-11a, (R,R)-11b, (R,R)-12a or (R,R)-12b)-(R)-BINAP-RuCl$_2$, it has been shown that the suitable solvent is a mixture of i-PrOH/DMF/ (1/1, v/v). If i-PrOH is employed alone, hydrogenation does not take place due to little swelling of the polymer catalyst. (Scheme 2, Ar = Ph). Moreover, a lower degree of crosslinking gave better reactivity; higher ees and conversion are obtained (Table 1).

The best results are observed if xylBINAP is used instead of BINAP (ie 93% ee, 100% yield in the presence of 10a). The immobilized (S,S)-10a-(R)-BINAP-RuCl$_2$ can be reused at least four times without loss of activity and selectivity (from the first to the fourth reuse ee is about 73-74% with 100% yield) [12].

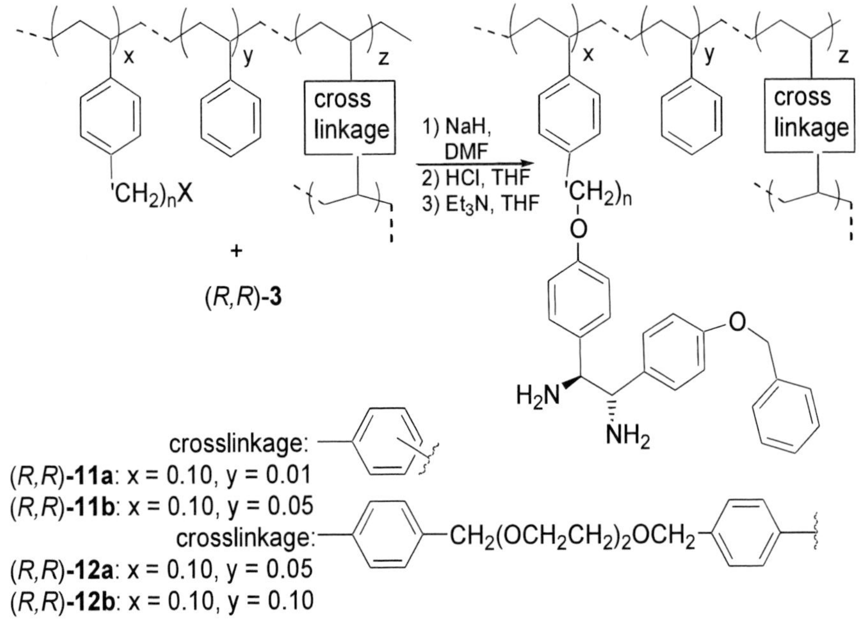
x
y
z
cross linkage
'CH2)nX
+
(R,R)-3
1) NaH, DMF
2) HCl, THF
3) Et3N, THF
'CH2)n
O
H2N
NH2
crosslinkage:
(R,R)-11a: x = 0.10, y = 0.01
(R,R)-11b: x = 0.10, y = 0.05
crosslinkage:
CH2(OCH2CH2)2OCH2
(R,R)-12a: x = 0.10, y = 0.05
(R,R)-12b: x = 0.10, y = 0.10

Scheme 5.

Table 1. Enantioselective hydrogenation of acetophenone with diamine polymer-supported-BINAP-$RuCl_2$

Ligand-BINAP-$RuCl_2$	Yield (%)	Ee (%)	Ref.
10a[a]	100%	73%	[12]
10b[a]	39%	68%	[12]
11a[b]	91%	70%	[13]
11b[b]	46%	67%	[13]
12a[b]	>99%	23%	[13]
12b[b]	70%	65%	[13]

[a] Diamine/BINAP/$RuCl_2$/ketone/*t*-BuOK : 1/1/1/200/20.

[b] Diamine/BINAP/$RuCl_2$/ketone/*t*-BuOK : 2/1/1/1000/20.

Enantioselectivities obtained with cocatalyst DPEN modified by a vinylphenyl group copolymerized under radical conditions (polymers **13** [14] and **14, 15** (Schemes 6 and 7) [15] and **16** (Scheme 8) [13] also exhibited almost the same level as those obtained from the low-molecular-weight counterpart in solution system (73-80% and 80% ee respectively with conversions always >99% for polymer containing less than 5% DVB).

Catalytic systems derived from **13** and **14, 15** and **16a** have successfully been recycled from four (for **13** [14]), **14** [15] and **15** [15]) to sixteen times (for **16a** [13]) without any change of activity and selectivity. They were also tested with other aromatic ketones and high yields and enantioselectivities were observed (some examples are presented Table 2).

For the hydrogenation of the racemic α-(N-benzoyl-N-methylamino) propiophenone, (Scheme 9) polymers containing N-substituted DPEN were prepared by radical copolymerization of (R,R)-**9** (Scheme 3) with various crosslinking agents (Scheme 10) [16].

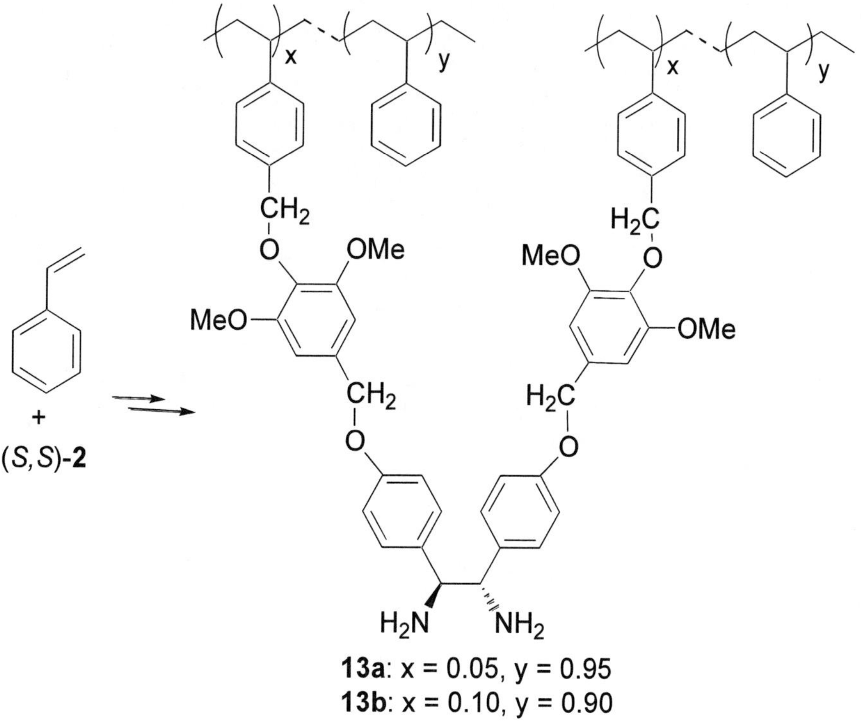

13a: x = 0.05, y = 0.95
13b: x = 0.10, y = 0.90

Scheme 6.

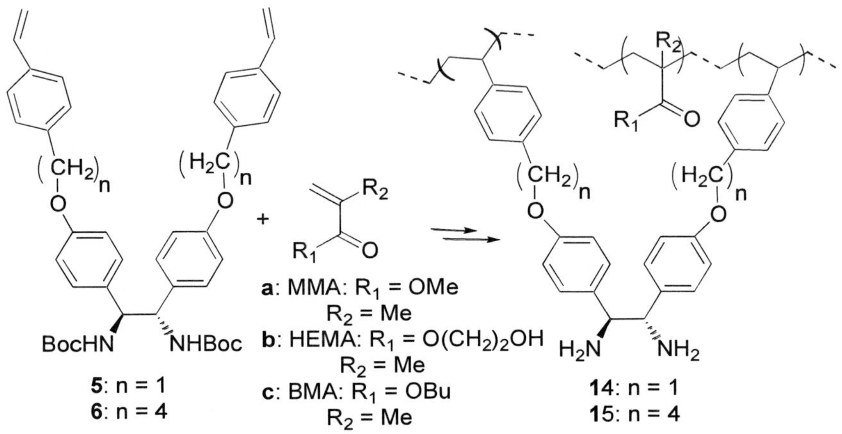

Scheme 7.

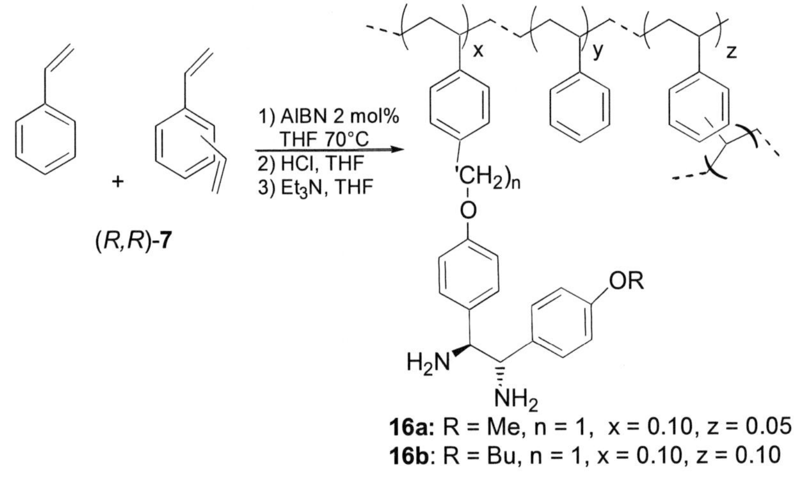

Scheme 8.

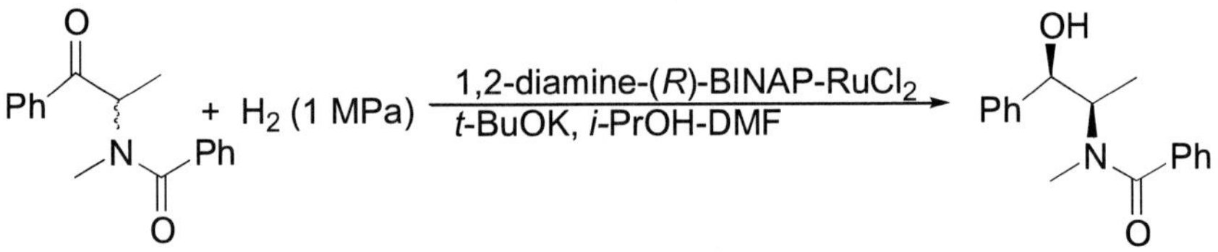

Scheme 9.

Table 2. Hydrogenation of several ketones with polymeric diamine/BINAP/RuCl$_2$ catalyst

Ketone	Diamine	Time (h)	Conv. (%)	Ee (%)	ref.
acetophenone	13a[a] or 13b[a]	18	100	77 (R)	[14]
	14a[b]	5	>99	78 (R)	[15]
	15c[b]	5	>99	79 (R)	[15]
propiophenone	13a[a]	4	100	82 (R)	[14]
	14a[b]	12	>99	84 (R)	[15]
	15c[b]	12	>99	83 (R)	[15]
	16a	24	99	82	[13]
1'-acetonaphtone	13a[a]	18	100	96 (R)	[14]
	14a[b]	12	>99	96 (R)	[15]
	15c[b]	12	>99	96 (R)	[15]
	16a[c]	24	>99	97	[13]

[a] Diamine/BINAP/RuCl$_2$/ketone/t-BuOK : 2/1/1/1000/20.
[b] Diamine/BINAP/RuCl$_2$/ketone/t-BuOK : 2/1/1/200/20.
[c] Diamine/BINAP/RuCl$_2$/ketone/t-BuOK : 1/1/1/200/20.

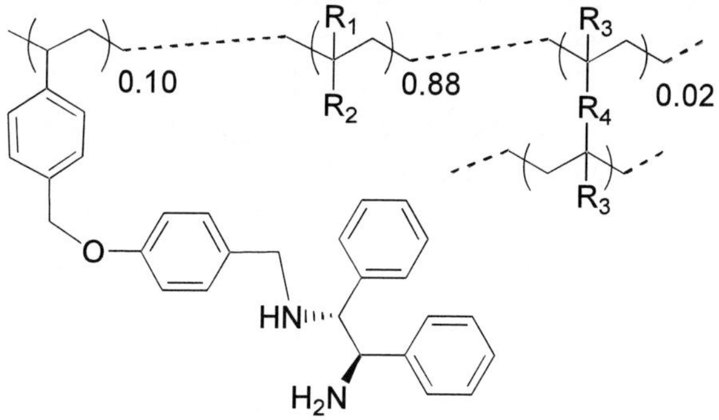

17a: R$_1$ = R$_3$ = H, R$_2$ = Ph, R$_4$ = C$_6$H$_4$
17b: R$_1$ = H, R$_2$ = Ph, R$_3$ = Me, R$_4$ = CO$_2$CH$_2$CH$_2$OCO
17c: R$_1$ = Me, R$_2$ = CO$_2$CH$_2$CH$_2$OH, R$_3$ = Me, R$_4$ = CO$_2$CH$_2$CH$_2$OCO
17d: R$_1$ = Me, R$_2$ = CO$_2$CH$_2$CH$_2$OH, R$_3$ = H, R$_4$ = C$_6$H$_4$
17e: R$_1$ = R$_3$ = Me, R$_2$ = CO$_2$Me, R$_4$ = CO$_2$CH$_2$CH$_2$OCO
17f: R$_1$ = H, R$_2$ = CONHi-PrMe, R$_3$ = Me, R$_4$ = CO$_2$CH$_2$CH$_2$OCO
17g: R$_1$ = H, R$_2$ = C$_6$H$_4$CH$_2$OH, R$_3$ = Me, R$_4$ = CO$_2$CH$_2$CH$_2$OCO

Scheme 10.

For the catalytic reduction, the solvent of choice was the mixture 1/1 (v/v) of i-PrOH/DMF (conditions for good swelling of the polymeric ligands **17**). The polarity of the crosslinking agent has a direct influence on the results of the catalysis. The precatalyst obtained with **17a** produced a low conversion (7%) and an excellent diastereoselection (>99%) for an enantiomeric excess of 61%. A slight improvement was observed for **17b** formed with EGDMA a polar crosslinker (conversion: 10%, ee: 88%). Compared to them **17e** present a better conversion (89%) with >99% of de and ee. The precatalysts prepared from **17c**, **17d**, **17f**, **17g** from better hydrophilic polymer support have exhibited excellent catalytic performance for this dynamic kinetic resolution of racemic α-amide ketone *via* hydrogenation to yield the corresponding *syn* β-amide alcohol (100% conversion, de >99% and ee ranging from 90 to >99%). In addition, the polymer-supported catalyst could be reused several times without loss of catalytic activity.

2.2. HYDROGENATION TRANSFER REDUCTION

Hydrogenation transfer reduction (HTR) is defined as a reduction of mainly C=O and C=N bonds employing an hydrogen donor (DH_2) such as cyclohexene, cyclohexadiene, alcohols, formic acid or hydrazine, in the presence of a catalyst, avoiding all the risks inherent to molecular hydrogen (Scheme 11). For this reason and for a procedural simplicity, HTR can be regarded as an alternative for the asymmetric hydrogenation [17]. For HTR, the first heterogeneous catalytic systems were formed with homogeneous-supported phosphines, which were also oxygen sensitive contrary to amine ligands. Moreover, nitrogen-containing ligands could be easily polymerized.

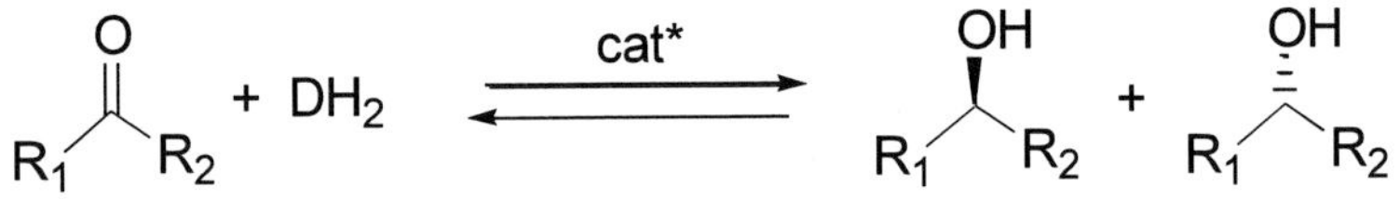

Scheme 11.

2.2.1. 1,2-Diamine and Derivatives

As for hydrogenation, most of the chiral ligands involved in heterogeneous HTR have been prepared starting from functionalized DPEN (Scheme 12).

In 1993, Lemaire found that *N,N*-dimethyl-1,2-diphenylethylenediamine (**18**) complexed with Rh could be employed as ligand for asymmetric HTR, leading to 67% ee and 100% for the reduction of acetophenone and methyl-phenyl glyoxalate respectively [18]. Due to the presence of the amino groups, this C_2 diamine could be easily transformed into diurea and dithiourea derivatives.

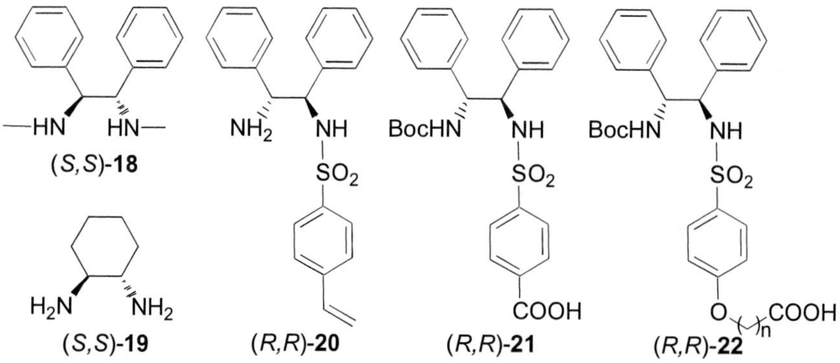

Scheme 12.

As good results for the asymmetric HTR of acetophenone were obtained (conversion 100% and 91% ee) with diurea [19], not only copolymerization of diamine **18** has been performed but 1,2-cyclohexyldiamine **19** was also used. Thus pseudo-C_2 polyamide **23**, polyureas **24, 24** and **26** or polythioureas **27** were prepared by polycondensation with diacid chloride, diisocyanate [20] or dithioisocyanate respectively (Scheme 12) [21]. With rhodium complexes the conversions varied from 22% to 100% and ee from 0% to 60% for HTR with acetophenone at 70°C, in the presence of [Rh(cod)Cl]$_2$ with diamine polymer unit (**23-26**)/Rh ratio of 10, 2-propanol and KOH. Polyamide **23** proved to be useless (only 22% conversion and 28% ee) contrary to polyurea **25** which presents similar ee to those observed with diamine **18** when (Rh(cod)Cl)$_2$ was employed as the catalytic precursor for HTR (ie 100% conversion for both of them and 55% and 59% ee respectively). Polyureas **24a, 24b, 25** and crosslinked **26** led to better conversions, 80, 50, 97 and 100% with respectively 0, 13, 39 and 60% ee. Moreover, the chiral crosslinked polyurea **26** presented a slight increase in enantioselectivity over the monomer analog **18** (55% ee and 94% conversion under similar conditions) and the reaction rate appeared to be even higher than in the homogeneous phase.

Catalytic system from **25** showed a capacity to recycle (Scheme 13) [20]. For thioureas, it has already been shown that under homogeneous conditions, they are better ligands than ureas, when using ruthenium complexes as precursors [22].

Thus, polythioureas were good candidates for the HTR reaction [21]. For polythioureas **27**, studies of the DP were performed and it had been showed that DP depends on the ratio of the reagents. During the polymerization, a guanidine moiety could be formed, stopping the chain growing. To avoid its formation, dithioisocyanate must be used in excess.

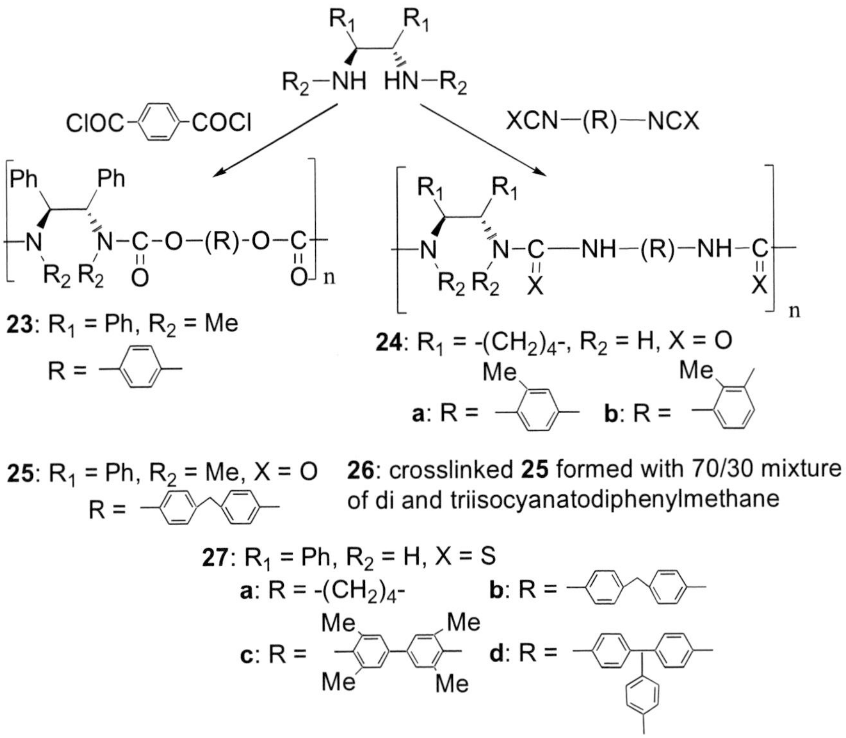

Scheme 13.

The HTR of various ketones was then performed in isopropanol with a substrate/ligand/t-BuOK/Ru ratio of 20/1.5/4/1. The chiral induction depended on the flexibility of the linker. The more flexible the linker, the less enantioselective the reaction: 31%, 70%, 70% and 65% ee for respectively **27a**, **27b**, **27c**, **27d** ruthenium complexes with conversion up to 90% for **27a**, **27b**, **27c** and only 47% for **27d**. For this latter ligand which was fully crosslinked, the decrease in conversion was due to the low accessibility of the catalytic site although a ligand/metal ratio of 8 had to be used.

The recovery and reusing of the **27c** polymeric catalyst has been performed with practically no loss of either catalytic activity or enantio-selectivity (Scheme 14). These results could be explained by the rigidity of the active site which is crucial for the selectivity and the stability of the catalytic system. Compared to the polyureas most of the polythioureas tested were more efficient.

Modification of 1,2-diphenylethylenediamine into *N*-*p*-toluenesulfonyl-1,2-diphenylethylenediamine (TsDPEN) led to one of the most selective ligand for the homogeneous HTR of ketones (97% ee for acetophenone) [23].

The heterogeneization of this ligand was first carried out with *N*-(vinylbenzene-*p*-sulfonyl)-1,2-diphenylethylenediamine (**20**), styrene and divinylbenzene as crosslinker (Scheme 15).

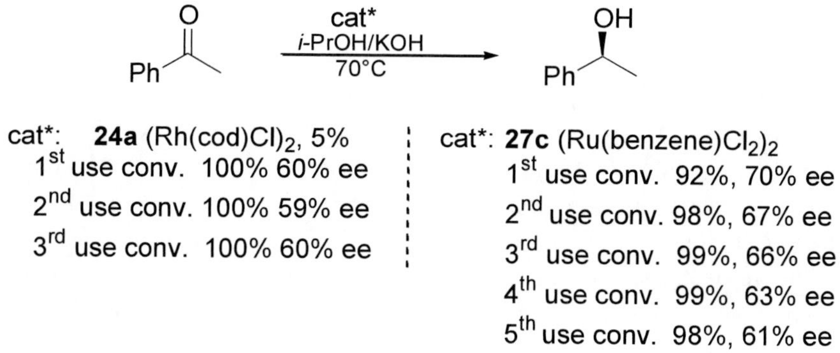
O
Ph
cat*
i-PrOH/KOH
70°C
OH
Ph
cat*: 24a (Rh(cod)Cl)2, 5%
1st use conv. 100% 60% ee
2nd use conv. 100% 59% ee
3rd use conv. 100% 60% ee
cat*: 27c (Ru(benzene)Cl2)2
1st use conv. 92%, 70% ee
2nd use conv. 98%, 67% ee
3rd use conv. 99%, 66% ee
4th use conv. 99%, 63% ee
5th use conv. 98%, 61% ee

Scheme 14.

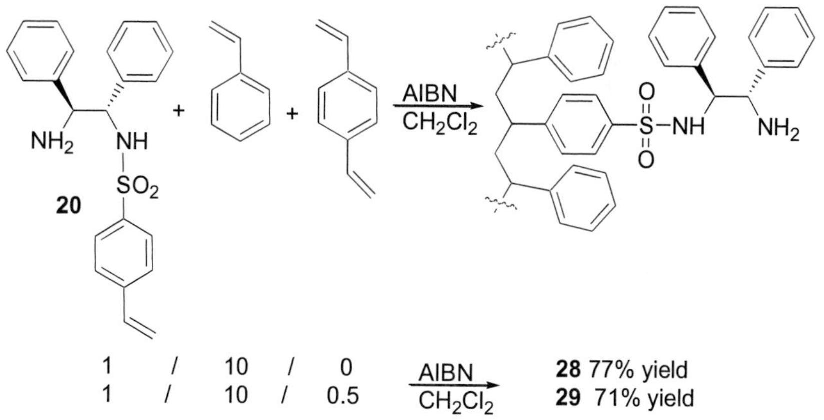
NH2
NH
SO2
20
+
+
AIBN
CH2Cl2
O
S
NH
O
NH2
1 / 10 / 0
1 / 10 / 0.5
AIBN
CH2Cl2
28 77% yield
29 71% yield

Scheme 15.

When tested as ligands for the Ir^I and Ru^{II} catalyzed HTR [24] of acetophenone, in the presence of 2.5% of metal precursor and 2-propanol/base, at 70°C, Ir complexes led to higher ee (**28**: 92% and for the crosslinked **29**: 94%) compared to the homogeneous analog for which only 75% of conversion and 89% ee were obtained. Nevertheless, the Ir catalyst was more efficient than ruthenium (Ir: 92% ee, Ru: 64% ee crosslinked Ir: 94% ee, crosslinked Ru: 84% ee), but for Ru the conversions were of about 20% in both cases. On the other hand the homogeneous Ir complex analog led to only 75% conversion at 89% ee. This difference could be explained by the formation of chiral microenvironments upon polymerization and stabilization of the reactive Ir complex.

Attempts to reuse the catalysts showed that the one containing Ru which is less selective is more stable upon reuse than the Ir derivative.

Itsuno [25] has also shown that polymer-supported DPEN monosulfonamides containing sulfonated pendent group (Scheme 16) are able to catalyze the HTR reduction of ketones in water with sodium formiate as hydrogen donor (S/C = 100). However, TsDPEN immobilized on polystyrene crosslinked or not, polymer **30** and **31** respectively, shrank in water. Sodium *p*-styrene sulfonate was copolymerized with chiral *N*-(vinylbenzene-*p*-sulfonyl)-DPEN (**20**) under radical polymerization conditions with or without DVB leading respectively to ligand **32** and **33**. Control of the balance hydrophilicity/hydrophobicity of the polymer support is carried out by changing the salt from Na^+ to quaternary ammonium. All of these polymers swelled in water, and their respective ruthenium, rhodium or iridium complexes were prepared. Compared to sodium salt polymer-supported catalyst from **32a** and **33a**, ammonium supported catalyst from **32b** and **33b** have produced the best results in terms of activity and selectivity for the reaction in aqueous media. Moreover, almost the same activity was observed for the reduction of acetophenone with the ammonium non crosslinked catalyst **32b** prepared from $[RhCl_2Cp]_2$ and $[RuCl_2(p\text{-cymene})]_2$ (100% and 99% conversion; 98% and 98% ee respectively) while the catalyst from $[IrCl_2Cp]_2$ decreased both reactivity and enantioselectivity (77% conversion, 89% ee).

Enantioselectivities of 97-98% were obtained in five recycle runs by using (*R,R*)-**33b** derived catalyst.

Earlier, Bayston [26] has synthesized the polymer **34** (Scheme 17) starting from the aminostyrene Merrifield-type polymer. By means of ruthenium precursor, 88% conversion with 91% ee was obtained for reduction of acetophenone. Compared with the preceding ligand **29** complexed with ruthenium, better conversion and enantioselectivities are shown; these latter results were comparable with the Ir catalyst.

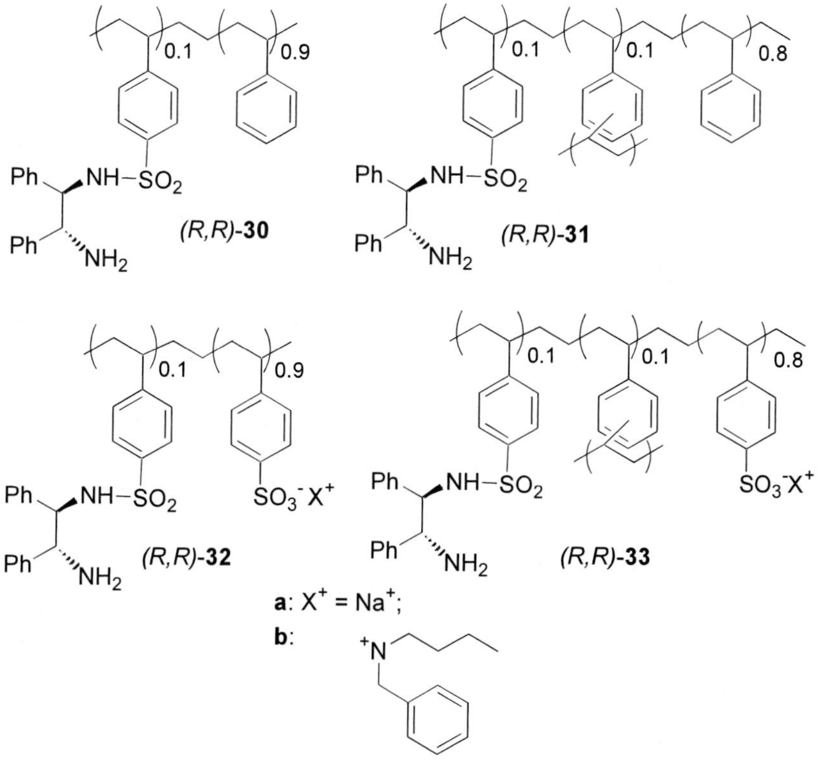

Scheme 16.

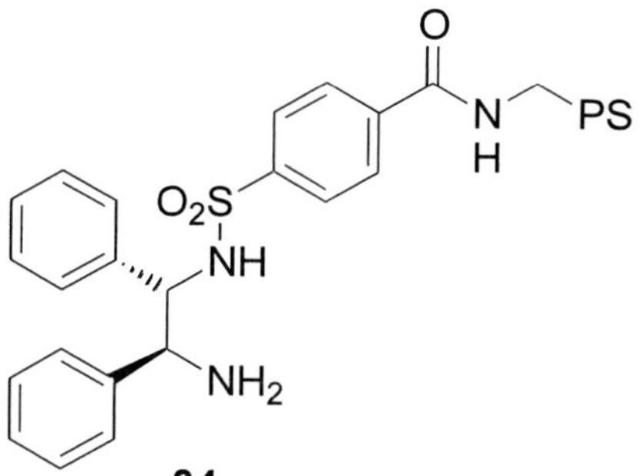

34

Scheme 17.

A similar heterogeneization of TsDPEN was made by Wang [27] with *N*-(carboxy, carbomethoxy or carboethoxy) (benzene -*p*-sulfonyl)-1,2-diphenylethylenediamine (**21** and **22** scheme 12). The acid function permitted the immobilization of the ligand to an aminomethylated polystyrene *via* an amide function (Scheme 18).

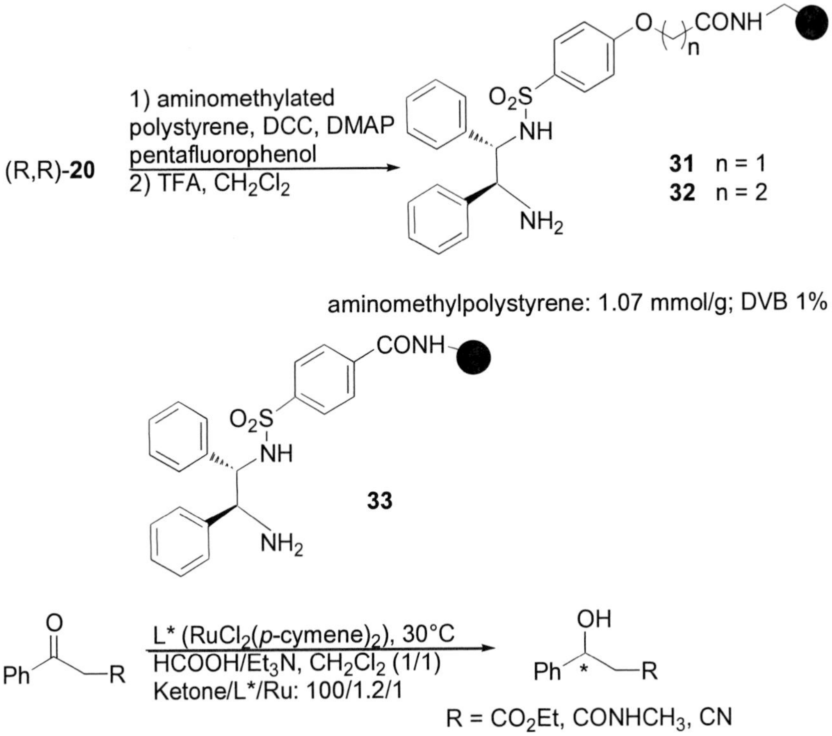

Scheme 18.

Ruthenium complexes of **31**, **32** and **33** using the azeotrope triethylamine/formic acid as hydrogen donor permitted the reduction of β-keto ester, amide and nitrile with high conversion (> 95%) and ee (> 90%) (Scheme 18). The electronic character and the spacer length between the polystyrene part and the benzene ring had very little effect on the reduction outcome, for a same substrate conversion and ee are similar. Here also, it has been shown that the catalyst formed with ligand **32** could be reused at least three times without loss of activity and enantioselectivity.

Xiao proposed a poly(ethylene glycol)-supported ligand (**34**) in order to perform the HTR of aromatic ketones in water, using as hydrogen donor, the azeotrope triethylamine/formic acid [28] or sodium formate/water (Scheme 19) [29]. These ligands were formed by immobilization of TsDPEN by functionalization of the phenyl group by an hydroxy group in the *meta* position followed by a nucleophilic reaction with a monomethylether mesylate PEG (Mn 2000) (Scheme 19). The Ru-catalyst formed was soluble under the conditions of the reaction. It could be isolated from the reaction mixture by precipitation, though addition of a low polarity solvent, allowing recycling. Up to 80% ee and up to 85% ee were achieved with all of the ketones tested using the azeotrope triethylamine/formic acid or sodium formate/water systems respectively. Nevertheless, convertion depended on the structure of the ketone and mainly on the position of the withdrawing group of the aromatic group (Cl, OMe, CF_3).

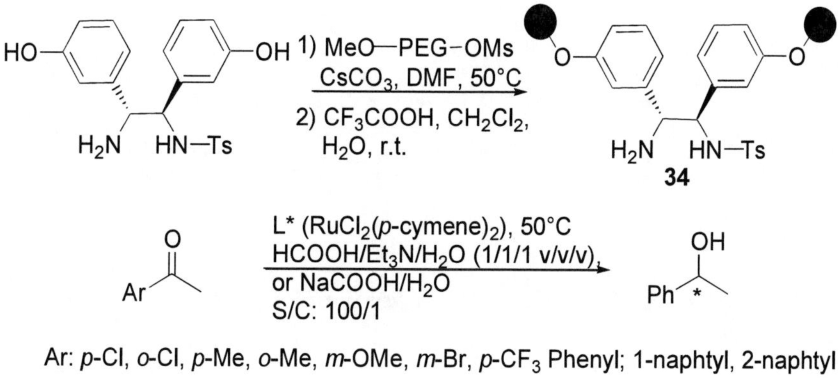

Ar: *p*-Cl, *o*-Cl, *p*-Me, *o*-Me, *m*-OMe, *m*-Br, *p*-CF$_3$ Phenyl; 1-naphtyl, 2-naphtyl

Scheme 19.

For the sodium formate/water reduction system [29], several parameters were studied. A 20°C decrease of the temperature led to a slightly higher ee at a longer time. The lower the formate concentration, the slower the reduction is and the lower ee. Moreover, the reaction was not affected by the presence of a surfactant. The ligand itself probably acted as a phase transfer catalyst. The presence of an organic solvent, inducing a biphasic system, led to a slower reaction and a decrease in ee. Concerning the catalytic system stability, it appeared that in sodium formate/water medium, a possible decomposition occurred because catalytic recycle led quickly to loss of catalytic activity and ee, which was not the case when formate/water system was used: more than ten recycles with no loss of activity and enantioselectivity.

The same approach was involved in 2008 by Chan [30]. As an alternative method for attaching a PEG chain onto TsDPEN ligand, a PEG-750 (ie Mn =

750) was anchored at the *para*-position of an aryl sulphate group leading to polymer **35** in 51% overall yields starting from chiral DPEN (Scheme 20). Polymer 35 was examined in the Ru-catalyzed asymmetric HTR of acetophenone in water using HCOONa as the hydrogen donor at room temperature. Best ee (96%) was obtained with 5 equivalents of HCOONa and the high ratio of substrate to catalyst (S/C = 1000) led to slower reaction with a slight decrease in ee (90% ee). Other aryl ketones have been reduced. Compared with the precedent catalytic system formed with polymer **34**, similar results were obtained in term of enantioselectivity. Once again, the catalyst could be easily recovered and reused at least eight times.

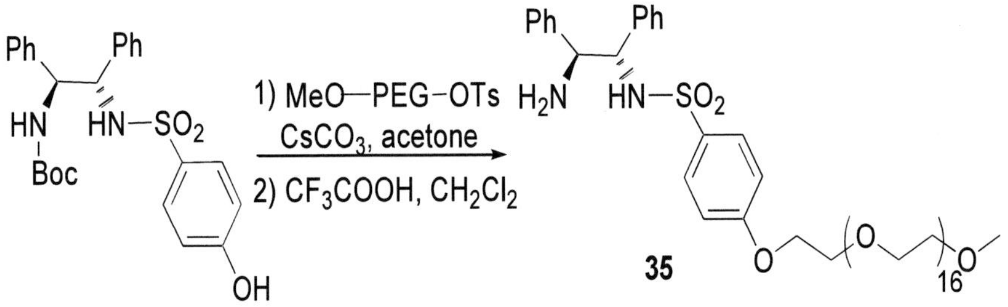

Scheme 20.

2.2.2. *Imprinting Technique for Hydrogen Transfer Reduction with 1,2-Diamines*

In 1995, Lemaire [32] reported the immobilized transfer hydrogenation catalyst using a tetramine-rhodium complex **35** in order to study the effect of molecular imprinting. The upper rhodium complex was formed by copolymerization with a diisocyanide in the presence of sodium (*S*)-phenylethanolate (PM) leading to the imprinted polymers **36a** and **36b** (Scheme 21). The PM is then removed by adding 2-propanol. The resulting polymer as well as the non-templated polymer were tested in the transfer hydrogenation of acetophenone and phenylethylketone in the presence of 5 mol % of Rh-polymer with KOH/[Rh] ratio of 4 at 60°C. Results are presented in Table 3.

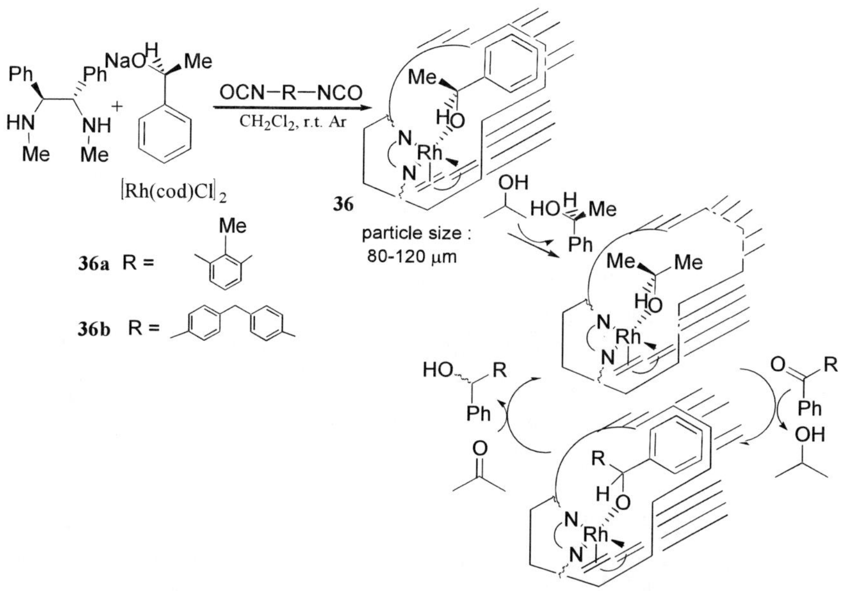

Scheme 21.

In homogeneous catalysis using chiral diamine **18** complexed with Rh, the acetophenone was reduced quantitatively with 55% ee, in 7 days. In the case of polymerized complex **36a**, acetophenone reduction leads to 33% ee and with its templated analog 43% ee. With **36b**, an increase of about 20% ee is observed between polymerized and templated ligand. These increases in ee were ascribed to a favourable molecular imprinting effect of the PM, creating chiral pockets within the polymer network.

Table 3. Molecular imprinting effect in reduction by using Rh catalyst with templated and non-templated polymers 35a and 35b

Substrate	Chiral polymer-Rh catalyst	Inductor configuration	Conv. (%)	Time (days)	Ee(%) config.
acetophenone	**36a** polymerized	(*R,R*)	44	1	33 (*S*)
acetophenone	**36a** (*S*) templated	(*R,R*)	42	1	43 (*S*)
acetophenone	**36b** polymerized	(*S,S*)	98	1	25 (*R*)
acetophenone	**36b** (*R*) templated	(*S,S*)	98	1	43 (*R*)
phenylethylketone	**36b** polymerized	(*S,S*)	96	6	47 (*R*)
phenylethylketone	**36b** (*R*) templated	(*S,S*)	91	9	67 (*R*)

Nevertheless, in the case of sterically demanding substrates little or no reduction occured, indicating a substrate stereospecificity different from that observed in homogeneous catalysis or when using catalyst precursor belonging to the polymer backbone [32]. For the reduction of acetophenone, crosslinked templated polymers were studied. Optimization of the crosslinking ratio led to the best compromise between activity and selectivity (70% ee for a crosslinking ratio of 50/50 of triisocyanate/diisocyanate) [33, 34].

This selectivity could be explained by a certain rigidity of the cavity, permitting good accessibility to the reaction sites.

This molecular imprinting technique was also used by Severin [35] [36] who proposed the synthesis of phosphonato complexes **37** (Scheme 22) in order to mimic the six membered cyclic transition structure suggested for HTR. This immobilized catalyst was prepared in a three step reaction (Scheme 22). In order to compare the effect of such polymeric organometallic transition state analog (TSA **38**), the polymer **40** was prepared (Scheme 22) by copolymerization of the non phosphonato complexed catalyst with ethylene glycol dimethacrylate (EGDMA) (Ru/EGDMA: 1/99). The ability of such polymers to catalyse the reduction of benzophenone was tested, using the azeotrope formic acid/Et$_3$N as hydrogen donor.

The molecular imprinted polymer **39a** was significantly more active than polymer **40a** (TOF (**40a**) = 51.4 h^{-1}, TOF (**39a**) = 16.5 h^{-1}). With this enhancement of the reaction rate using the imprinting polymer, substancially higher specificity for benzophenone was observed when completion of HTR of an equimolar amont of benzophenone and a ketonic cosubstrate was performed. This behaviour was slightly larger for aliphatic ketones as compared to aromatic ketones. These results were consistent with the activity and selectivity using organometallic TSA as template with an imprinting technique. A similar study was made with the polymer including rhodium complexes coordinated with phosphinato ligands (polymer **38b**) and the imprinted polymer **39b**. After 6 hours, the reduction of acetophenone led to 81% yields and 95% ee, and ruthenium complexes gave up to 70% ee for the reduction of the ethyl phenyl ketone [37].

Due to the structurally defined transition state analogs, included in a polymer, a shape-selective cavity close to the catalytic active center was formed. This cavity is highly selective and an enhancement of the activity was observed

Furthermore, it has also been shown that activity and selectivity depended on how the metal complex was attached to the polymer backbone: a rigid connection by two styrene side chains is superior [36].

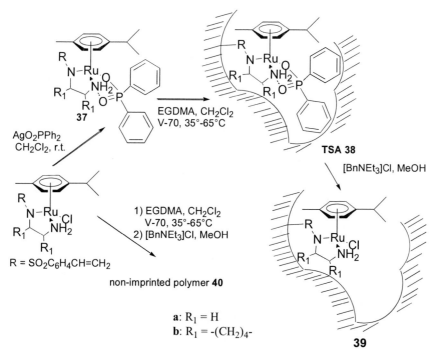

EGDMA, CH₂Cl₂
V-70, 35°-65°C

AgO₂PPh₂
CH₂Cl₂, r.t.

1) EGDMA, CH₂Cl₂
V-70, 35°-65°C
2) [BnNEt₃]Cl, MeOH

R = SO₂C₆H₄CH=CH₂

non-imprinted polymer **40**

a: R₁ = H
b: R₁ = -(CH₂)₄-

Scheme 22.

2.2.3. Aminoalcohol as Ligand

Chiral β-amino alcohols can also be used for homogeneous HTR of ketones and are known to be effective for reduction of aromatic ketones [38] but, only few examples of organic polymer-supported ligands have been reported.

Recently, poly((*S*)- glycidylmethacrylate- *co*- ethyleneglycol dimethacrylate) (**41**) (poly ((*S*)-GMA-*co*-EGDMA)), [39][40] (poly ((*S*)-PhGMA-*co*-EGDMA)) (**42**), [41] poly ((*S*)-glycidylmethacrylate-*co*-divinyl-benzene (poly ((*S*)-GMA-*co*-DVB)) (**43**) [41] [42] and ((*S*)-thiiranylmethyl-methacrylate-*co*-ethyleneglycol dimethacrylate) (**44**) [42] different precursor polymers were synthesized. They were formed by radical suspension polymerization of enantiopure (*S*)-GMA (30 wt % for **41a** or 70 wt % for **41b**) with EGDMA or DVB (70 wt % for **43**), or (*S*)-PhGMA (40 wt % for **42**) with EGDMA or (*S*)-TMA (30 wt % for **45**) with EGDMA in the presence of AIBN, with a mixture of cyclohexanol/dodecanol (91/9 wt/wt) and 2% of aqueous polyvinylpyrrolidone (PVP) as stabilizer (Scheme 23). Chiral beads of **41** and **43** presented respectively a specific surface area of 82 and 275 m^2/g, and in both the cases, 2.11 mmol/g of epoxyde function. A specific area of 50 m^2/g with 4.92 mmol/g of epoxyde function, a specific area of 35 m^2/g with 1.83 mmol/g of epoxyde function and a specific area of 92 m^2/g with 1.72 mmol/g of epoxyde function were respectively observed with **42, 44** and **45**. The subsequent epoxide or episufide ring opening with diverse amines led to the corresponding polyaminoalcohols and polyaminothiols. Their ruthenium complexes were applied to the HTR of benzophenone (Scheme 23).

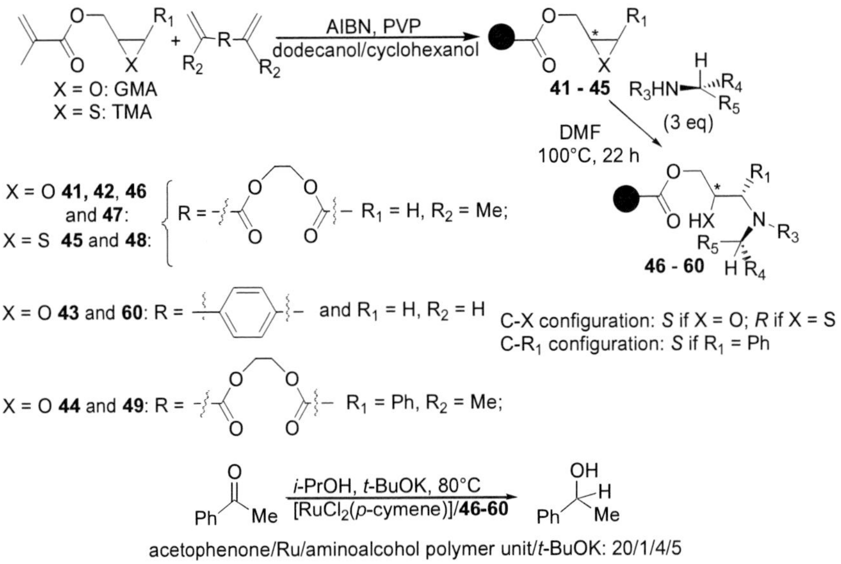

Scheme 23.

It is noteworthy that the nature of the amine of the aminoalcohol played a crucial role. The best results in terms of conversion and enantioselection are obtained with polyaminoalcohols and polyaminothiols derived from methylamine and benzylamine (Table 4) [39].

It appeared that the efficiency of the catalyst depends on the nature and the proportion of the crosslinking agent (catalyst presenting 70% of EGDMA was more efficient than this with 30% of EGDMA and for DVB a decrease in conversion and ee was observed). It depended also on the specific surface, the nature of the ligand aminoalcohol or aminothiol and on the steric hindrance (aminoalcohol and ligands from GMA compared with PhGMA were more efficient) (Table 4) [41]. Moreover, attempts to recycle ruthenium complex of

46a led to a marked decrease in activity from 94 to 27% and lowered selectivity from 70 to 54%.

Wills [43] has tested a copolymer formed by 7/3 methylmethacrylate and hydroxyethylmethacrylate soluble in a range of organic solvent as backbone of a norephedrine pendent polymer ligand, an aminoalcohol ligand selected for its good results in HTR of ketones when complexed with Ru^{II} (Scheme 24).

Table 4. HTR of acetophenone using several aminoalcohols and aminothiols (L*/Ru/Ketone/t-BuOH: 4/1/20/5, 80°C)

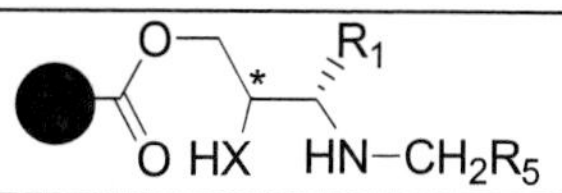

X C-X config.	R$_1$ C-R$_1$ config.	R$_5$	ligand	functionality (mmol/g)	time (h)	Conv. (%)	ee (%) config.
O (S)	H	Ph	**46a**	1,46	3	94	71 (R)
O(S)	H	H	**46b**	0,85	1	95	65 (R)
O (S)	H	Ph	**47a**	2.90	72	51	57 (R)
S (R)	H	Ph	**48a**	1.11	22	55	50 (S)
S (R)	H	H	**48b**	0,90	22	50	39 (S)
O (S)	Ph (S)	Ph	**49a**	0.65	22	61	21 (R)
O (S)	Ph (S)	H	**49b**	0.51	72	<5	29 (R)
O (S)	H	Ph	**60a**	1.28	22	58	44 (R)

(Table header figure caption: "Phenylethanol")

With ephedrine supported ligand **51**, the best results were obtained for the HTR of acetophenone when 0.44 mol% of catalyst in dichloromethane are used (20% yield, 77% ee). Evaluation of ligand **52** for this reaction with 2 mol% of catalyst gave up to 85% yield with 81% ee. Three additions of acetophenone showed a deterioration of the catalytic system.

Few attemps have been made by Masuda to utilize the main chain chirality of polymers [44]. They reported the synthesis of an L-threonine-based helical poly(N-propargylamide)-Ru complex (**53**-Ru-complex) and evaluation of its catalytic activity in the asymmetric HTR of ketones (Scheme 25). The corresponding alcohols were obtained in 23-48% yields and the ee values were 12-36%; the highest ee was achieved in the reduction of 2-acetonaphtone. This reaction was achieved in homogeneous conditions with **54**-Ru-complex and acetophenone led to phenylethanol with no ee. This suggested that the helical structure was effective to enhance the catalytic ability for chiral induction.

2.2.4. Dialdimine Ligand as Ligand

One example of the use of heterogeneized dialdimine ligand **55** (Scheme 26) was reported for the asymmetric HTR of acetophenone [45]. Enantioselectivities of up to 70% were obtained for a degree of crosslinking of 71%. The rise of the degree of crosslinking to 100% gave no significant difference contrary to a crosslinking less than 71% which gave a decrease in enantioselectivity.

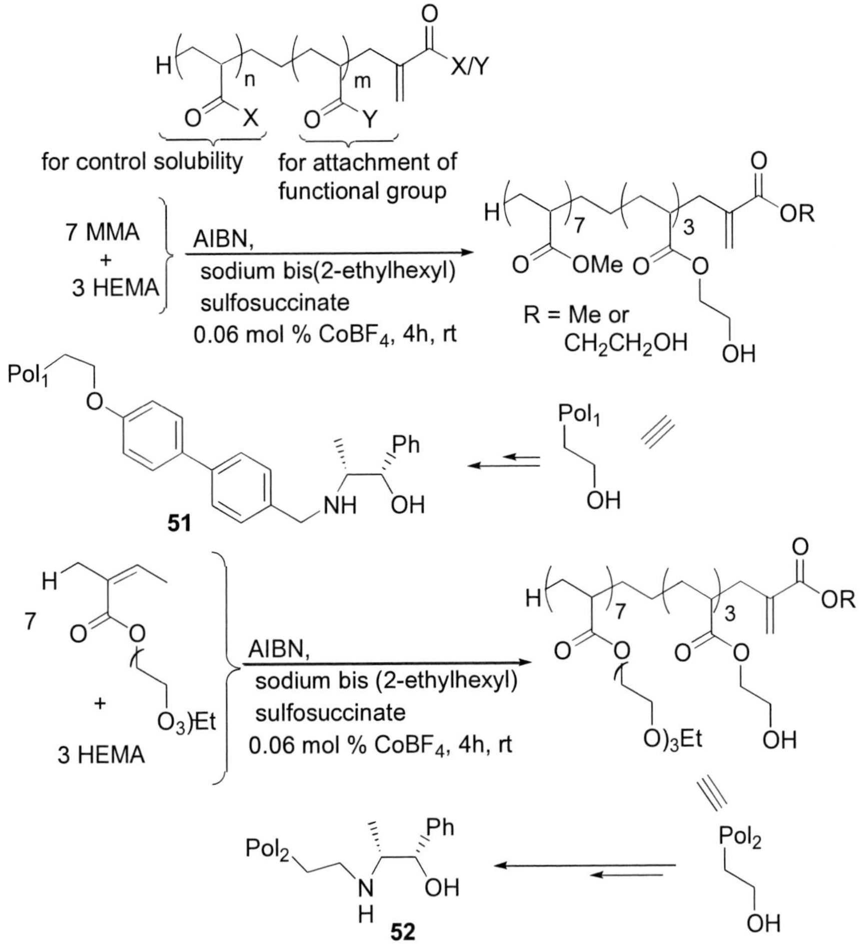

Scheme 24.

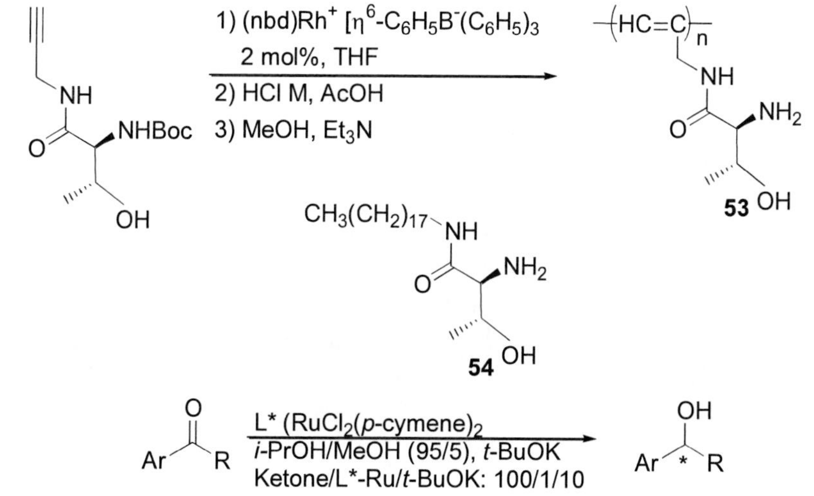

Scheme 25.

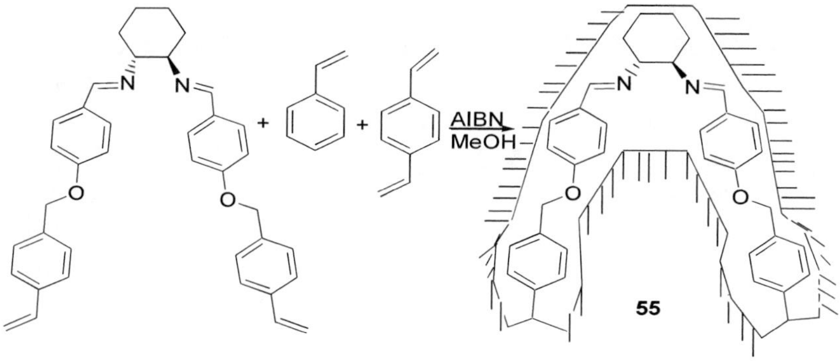

Scheme 26.

2.3. ENANTIOSELECTIVE HYDRIDE REDUCTION

The enantioselective hydride reduction of prochiral ketone is a well documented reaction which consists in the preparation of optically active alcohols. Thus, a large number of reducing agents and chiral catalysts have been developed.

In 1981, Itsuno [46] then Corey in 1987 [47] described the use of oxazaborolidine in the presence of an hydride for ketone reductions which is one of the most efficient system for this type of reaction. [48] The chiral ligands are mostly aminoalcohols such as chiral prolinol or ephedrine-type compounds.

Thus, from the beginning of the eighties, polymers containing chiral aminoalcohols and their derivatives were used as chiral auxilliaries for LAH [49], BH$_3$ [50] [51] or NaBH$_4$ [52].

One of the first experiment involved a continuous flow system using a crosslinked polystyrene gel supporting a chiral aminoalcohol **56** [53] (Scheme 27), prepared from L-tyrosine and 2% crosslinked chloromethylated poly-styrene gel [51]. The reducing agent is formed by a polymeric complex of amino-alcohol **56** and BH$_3$. Valerophenone and BH$_3$ are continuously added and passed through the polymeric complex. With this method the destruction of the polymeric gel was not observed (no stirring) and large quantities of the corresponding optically active alcohol could be formed. With valerophenone, 83% to 91% ee are obtained.

With the soluble polymer-supported catalyst **57** (Scheme 27), the reduction of acetophenone was performed in a continuous operated membrane reactor equipped with a nanofiltration membrane [54]. An enhancement of total TON from 10 to 560 (equivalent to 0.18 mol % catalyst for an average ee of 91% and an almost quantitative conversion) was observed. Excellent space-time yields of up to 1.4 kg per day were reached

More recently, several polymer versions of the efficient Corey-Bakshi-Shibata (CBS) catalyst have been prepared [47]. Different approaches have been used to bind the catalytic moieties to the polymer support: the diphenyl prolinol reacted with beads bearing benzene sulfonyl chloride residues (catalytic moieties bound to the nitrogen atom as part of a sulfonamide link), the diphenyl prolinol is bound to a polymer support or the aminoalcohol reacted with a polymer-supported boronic acid [55]. As in the last case the aminoalcohol is not supported, only the first two possibilities are reported.

In order to prepare polymer-supported CBS catalyst, monomeric precursors **59, 60** or **61** or bromides **62** and **63** have been synthesized (Scheme 28). The oxazaborolidine could be formed *in situ* or not before performing the hydride reduction of ketones.

Shore [56] synthesized a pendent (**64**) and a crosslinked (**65**) CBS catalyst in bead form from monomeric precursors **59a** and **60** respectively (Scheme 29) but also in monolith polymer form with a crosslinking level of 2% [57]. Monolith polymers were then cut to form CBS disks (**64D** and **65D**) roughly 2 mm in height and 5 mm in diameter.

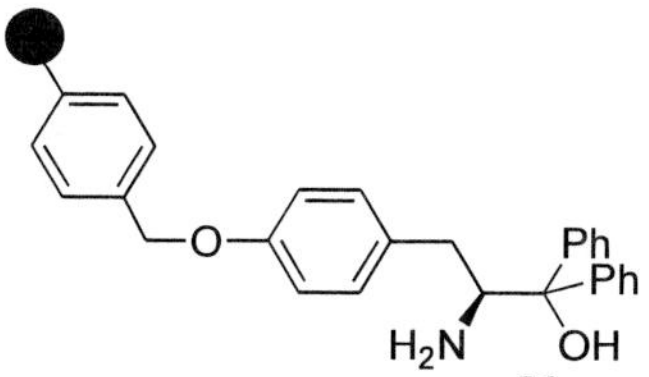

56

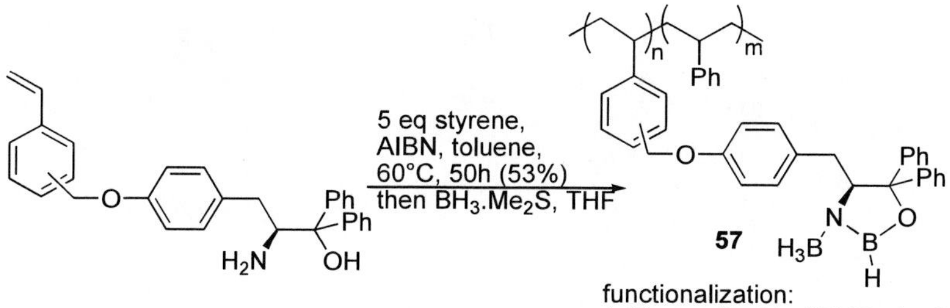

5 eq styrene,
AIBN, toluene,
60°C, 50h (53%)
then BH₃.Me₂S, THF

57

functionalization:
1.18 mmol/g, Mn 13800 g/mol

Scheme 27.

CBS ligands

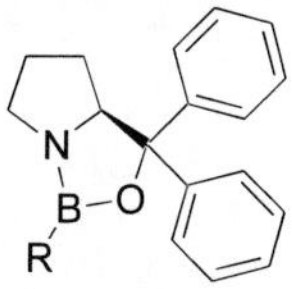

58a: R = H or **58b**: R = Me

CBS supported precursors

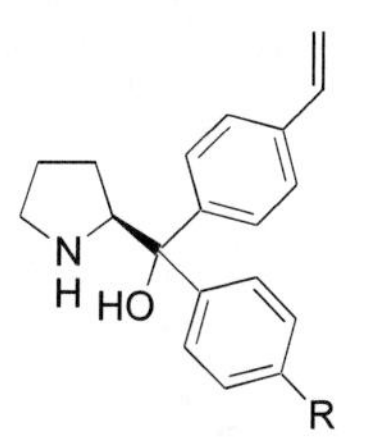

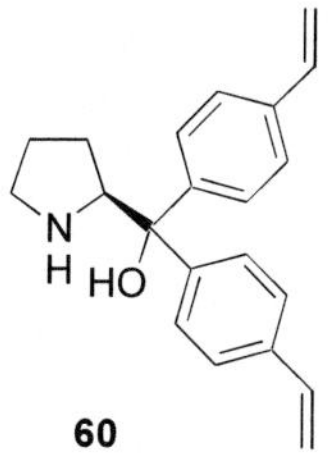

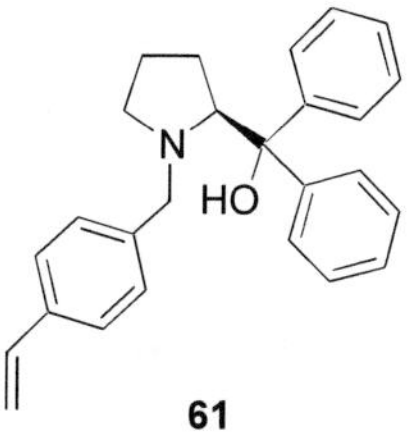

59a: R = p- i-Pr
59b: R = H

60

61

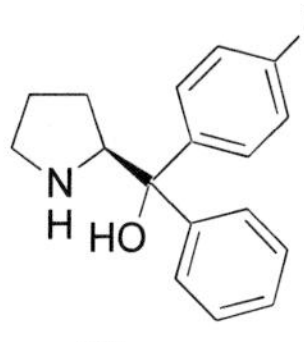

62

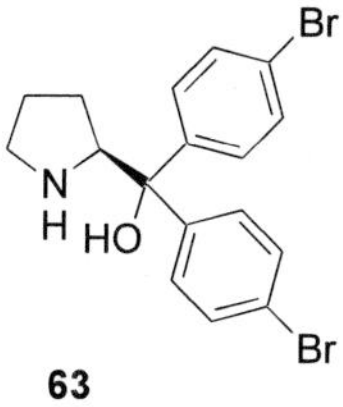

63

Scheme 28.

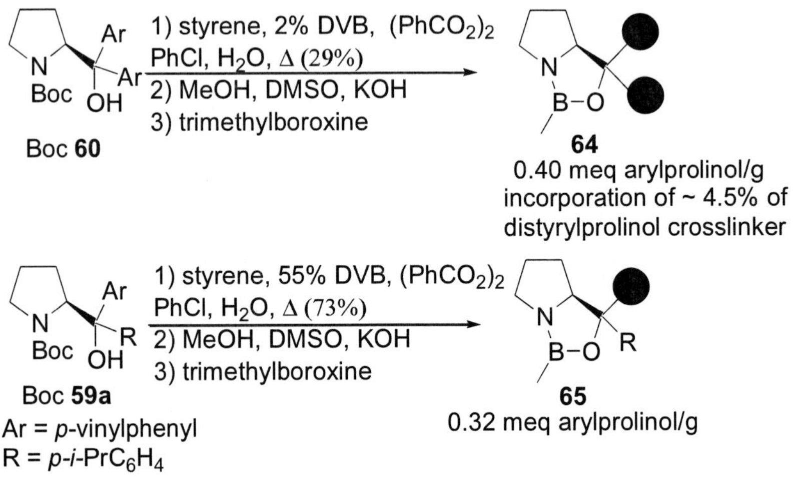

Scheme 29.

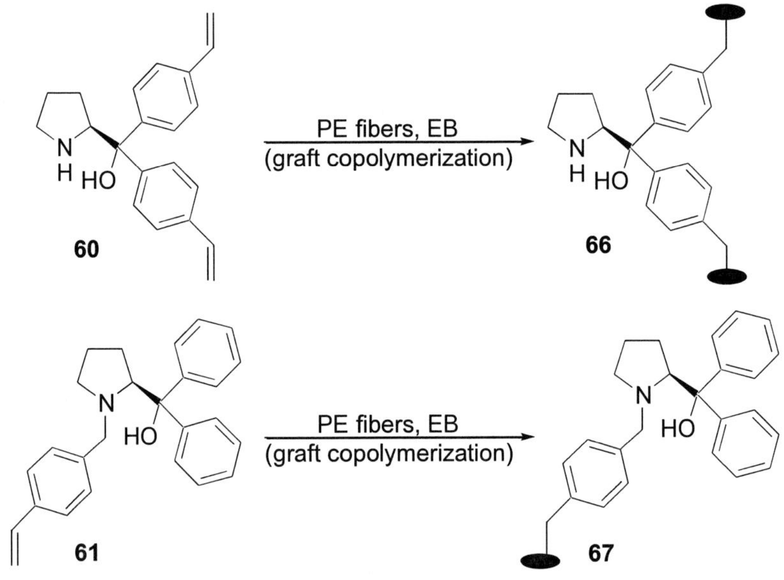

Scheme 30.

In 2004, Degni anchored chiral styrenic prolinol **60** and **61** on chemically inert polyethylene (PE) fibers by electron beam (EB) preirradiation, inducing graft copolymerization leading respectively to polymer **66** and **67** (Scheme 30) [58]. The corresponding oxazaborolidines were prepared *in situ* with borane from different sources: NaBH$_4$ in the presence of Me$_3$SiCl or BH$_3$.Et$_2$O and BH$_3$.SMe$_2$.

Involving a Suzuki reaction, Hodge prepared "chemical robust" polymer-supported diphenyl prolinol based catalysts **68** and **69** from **62** and **63** and polystyrene boronic acid prepared *via* a direct lithiation of 1% crosslinking gel-type polystyrene beads (2.21 mmg/mol of boronic acid residue) (Scheme 31) [59].

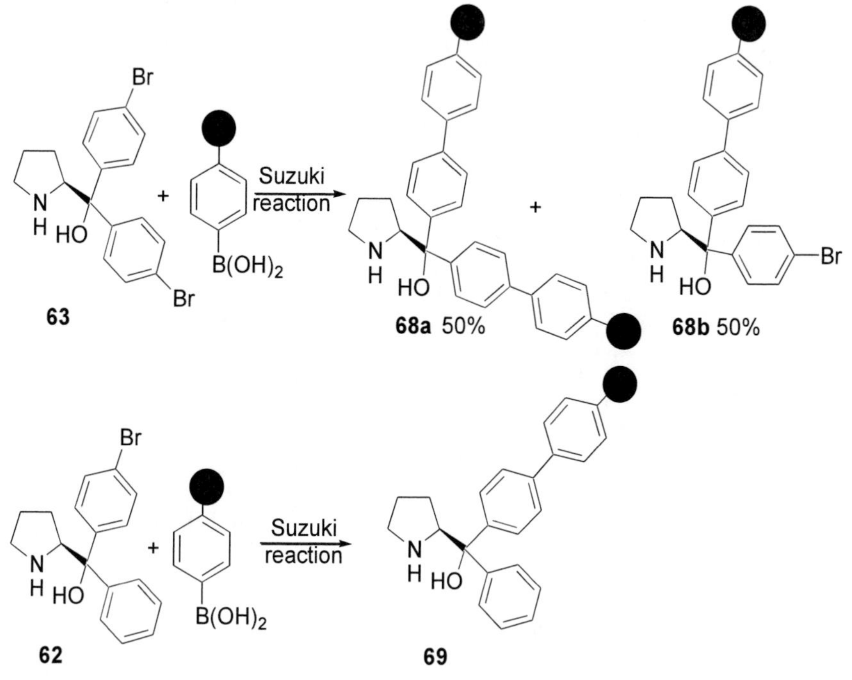

Suzuki reaction conditions: Na$_2$CO$_3$, (PPh$_3$)$_4$Pd, (CH$_2$OMe)$_2$

Scheme 31.

Comparative results for the reduction of acetophenone with these diverse oxazaborolidines are presented in Table 5.

In general, the crosslinked CBS (**64**, **64D** or **66**) afforded enantioselectivities almost identical to those of the CBS in homogeneous conditions (**58a** and **58b**) and superior to the pendent-linker equivalent ligand (*ie* respectively **65**, **65D** and **67**). These results could be ascribed to the superior swelling characteristics of the crosslinked catalyst in the solvent of the reaction which permits a better diffusion of the reagents into the catalyst

[56-58]. For polymer **67**, the weaker coordination with borane, due to the steric hindrance on the nitrogen atom, may be responsible of the slower catalysis. With **68** and **69** the results of the reaction were almost similar. After the 20th reuse, analysis of the polymeric mixture **68** showed that the overall weigh fell to 57% of its original value with an average loss per reaction of <3%. Moreover, nitrogen analysis was almost the same, suggesting that the catalyst groups were firmly bound and that the losses are probably due to physical attrition [59]. Concerning the reuse of the crosslinking polymer **64D**, the recycle failed. This suggested a disassembly of the oxazaborolidine and a structural breakdown of the polymer itself [57].

Table 5. reduction of acetophenone with several oxazaborolidines isolated or formed *in situ* from aminoalcohols and borane

Aminoalcohol or oxazaborolidine	Borane source	Mol %	T (°C)	Yield (%)	Ee (%)	Recycling (ee%)	Ref.
58a	BH$_3$.THF	5	r.t.	100	85	-	
64	BH$_3$.THF	5	r.t.	100	84	-	[56]
65	BH$_3$.THF	5	r.t.	100	62	-	[56]
64D	BH$_3$.THF	50	r.t.	97	99	3 (58)	[57]
65D	BH$_3$.THF	50	r.t.	100	63	-	[57]
58b	BH$_3$.SMe$_2$	10	45	99	98	-	
66	BH$_3$.SMe$_2$	10	45	90	95	1 (78)	[58]
67	BH$_3$.SMe$_2$	10	45	99	44	-	[58]
68	BH$_3$.SMe$_2$	30	22	> 95	94	16	[59]
69	BH$_3$.SMe$_2$	30	22	> 95	96	8	[59]

Ketones [60], β-functionalized ketones [61] [62] and *meso* cylic imide [63] have been enantioselectively reduced using polymer-supported chiral sulfonamides in the presence of boranes (NaBH$_4$/Me$_3$SiCl or BH$_3$.SMe$_2$), leading *in situ* to the corresponding oxazaborolidine polymer-supported chiral sulfonamide **70** (Scheme 32).

The reduction of unfunctionalized ketone was investigated at various conditions reactions: THF or toluene as solvent, 1% or 2% of crosslinking polystyrene resin and several degrees of functionalization.

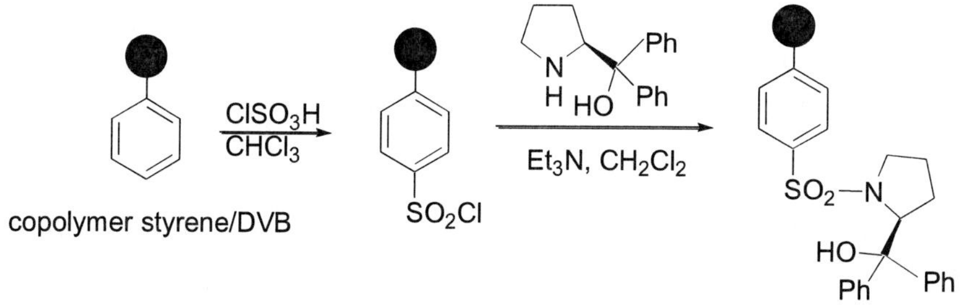

Scheme 32.

The best results in terms of enantioselection (92.5%) and conversion (95%) were obtained in reflux toluene, with a catalyst amount of 15% polymer crosslinking of 2% and an aminoalcohol loading of 2.29 mmol/g. With lower degree of functionalization and 1% crosslinking the ee was of about 10 to 20% lower. The catalyst has been reused at least five times with minimal loss of catalytic performance. It is noteworthy that under these conditions, homogeneous supported catalyst showed better results than homogeneous catalyst. The best conditions were also involved successfully for the reduction of other ketones.

With the same polymer-supported chiral sulfonamide **70** (200-400 mesh, 2% DVB, 2.29 mmol/g functionalization) the reduction of several β-ketonitriles [61] and β-ketosulfones [62] were tested

It is noteworthy that aromatic β-ketonitriles and β-ketosulfones were reduced in good enantioselectivity (85-96% ee). Electron density (substituant in the *para* position) had little effect on enantioselectivity. For aliphatic β-ketonitriles or sulfones, both chemical yield and ee were lower compared to those obtained with aromatic analogs.

Meso-cyclic imide precursor of *d*-biotin has successfully been reduced with BH$_3$-SMe$_2$ in the presence of polymer-supported sulfonamide catalyst **70** in THF under reflux for 6 hours. (Scheme 33) [63]. The enantioselective hydroxylactam was obtained in 91% yield with an ee > 98.5%. The recycling of the chiral supported-sulfonamide in the reduction of the *meso* precursor of *d*-biotin could be involved at least 5 times with non change in activity and enantioselectivity.

Recently, Wang [64] prepared by radical copolymerization a cinchona alkaloid copolymer: the methyl acrylate-*co*-quinine (PMA-QN (**71**)) (Scheme 34). Complexed with palladium(II), its catalytic activity in the heterogeneous catalytic reduction of aromatic ketones by sodium borohydride was studied. High yields in their corresponding alcohols are obtained but it is found that the efficiency of the catalyst depended on the nature of the solvent and the ketone which related to the accessibility of the catalytic active site. The optical yields in methanol and ethanol 95% were lower than in ethanol. This ability was attributed to a bad coordination between PMA-QN-PdCl$_2$ and sodium borohydride and a reaction rate which was very rapid. The stability of the chiral copolymer catalyst was studied *via* the study of the recycling efficiency. Only 2 wt% loss of metal was observed after repeated use (at least 5 times) with only a little decrease in enantioselectivity.

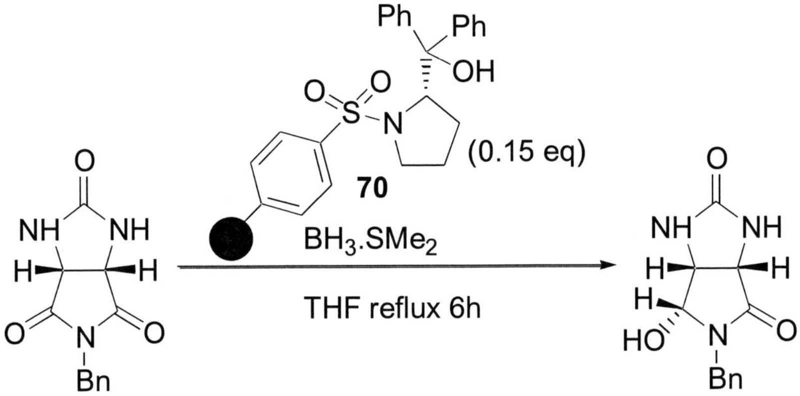

Scheme 33.

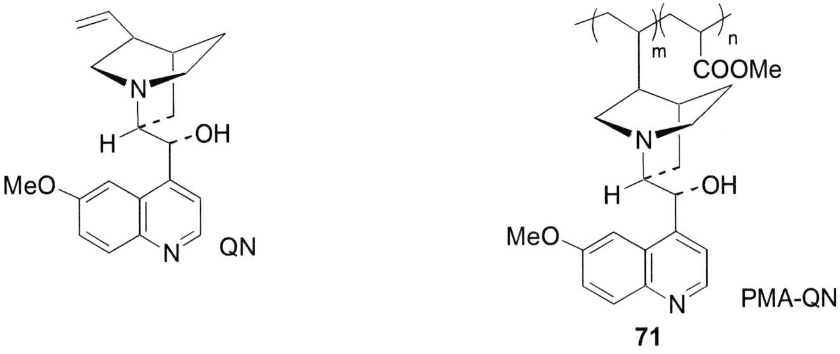

Scheme 34.

2.4. Hydrosilylation with Ligands as Pendent Group

Asymmetric hydrosilylation of ketone had been studied in homogeneous supported catalyst with nitrogen containing ligands and ony few examples have been reported. In 1998, Enders reported immobilized triazolium salts as precursors to chiral carbene-rhodium-catalyzed asymmetric hydrosilylation (Scheme 35). This Rh complex gave 24% ee compared to 17% ee for the homogeneous reaction (Scheme 34). The recycling of the solid supported catalyst **71** was successful with only slowly decreasing yields which were comparable to the yields of the homogeneous reactions with catalyst **72** [65].

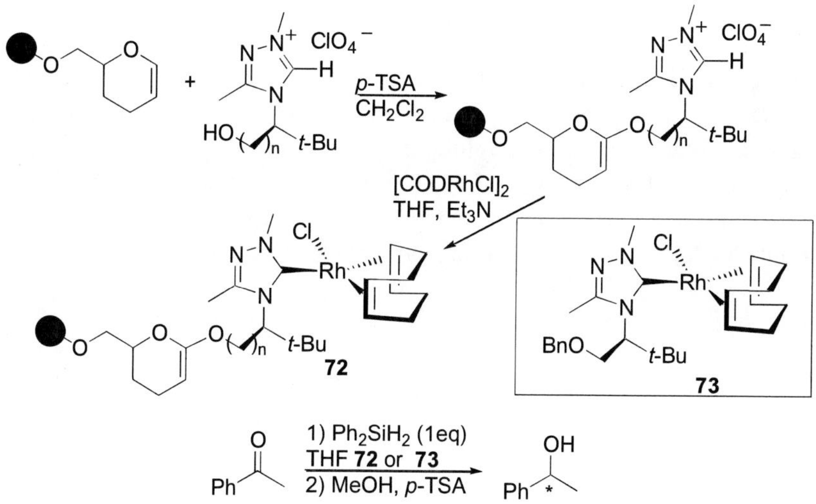

Scheme 35.

2.5. Reduction of C=N Bond

Although highly effective asymmetric reduction of carbonyl compounds has been extensively investigated, enantioselective reduction of imine derivatives to amines has been less studied, mainly in homogeneous supported catalysis [66].

With polymer containing nitrogen (2-amino-3-(p-hydroxyphenyl)-1,1-diphenylpropan-1-ol [67], 2-piperazinemethanol), Itsuno [68] has shown that in stoechiometric conditions, oxime ethers could be reduced with borane with excellent ee (up to 99%).

More recently, in organocatalysis, Kočovsky [69] has reported asymmetric reduction of imine with trichlorosilane catalyzed by an aminoacid(N-methylvaline)-derived formamide anchored to a polymer. Under the best conditions, with 15 mol% of the catalyst enantiomeric excesses were about 85% and the catalyst could be reuse.

Then, in 2009, Itsuno [70] utilized for HTR of N-benzylimine the polymer-immobilized chiral ligands 31, 33a and 33b (Scheme 16) as their Ru^{II} complexes. These complexes were already used to catalyze the HTR reduction of ketones in water [25].

The HTR was carried out in organic solvent (dichloromethane) and in water using as hydrogen donor respectively the azeotrope HCOOH/Et$_3$N and HCOONa. Catalytic system with ligand 31 was effective for the HTR of N-benzyl imines in organic solvent. The corresponding amines were formed in good yields and ee (92-96% yields and 84-88% ee). In contrast, Ru^{II} complexes obtained from amphiphilic polymer 33a and 33b were found to be effective for the HTR of cyclic imines in water (50-95% yields 86-94% ee). The catalytic activity, in water, seemed to be controlled by the hydrophilic-hydrophobic balance in a polymer-supported catalyst.

Chapter 3

CYCLOPROPANATION

Highly functionalized cyclopropanes are important building blocks for obtaining numerous natural compounds (terpenes, pheromones ....), and compounds which are of high value in biological (*ie*: agrochemical active species) and medicinal chemistry. A wide number of methods have emerged in order to produce optically active cyclopropanes [71].

From all of them, highly enantioselective cyclopropanes are formed by asymmetric catalytic insertion of carbene to prochiral olefins. This method, first employed by Noyori in 1966 [72] consists in the metal-catalyzed decomposition of substituted diazo compounds in the presence of various prochiral alkenes.

Under homogeneous conditions [73], various chiral Schiff base-Cu(II or I) or Ru^{II} complexes have been used as chiral catalyst precursors and from all of them the most efficient systems are formed with oxazolines and bis(oxazolines). Thus after their first use by Masamune [74], oxazoline-derived ligand have been extensively employed in homogeneous enantio-selective catalytic reaction particularly in cyclopropanation [75]. Thus nitrogen-anchoring ligands on organic polymers are mainly formed by immobilized oxazolines.

3.1. OXAZOLINES LIGANDS

Bis(oxazolines) (Box), azabis-oxazolines (azaBox) and pyridine Bis(oxazoline) (PyBox) with a C2 symmetry form the great majority of various oxazoline-derived ligands which are used in immobilized catalytic cyclopropanation reactions (Scheme 36). For these reactions, immobilized ruthenium and copper complexes which are able to form carbene intermediates have been tested. Most of the time, the reaction involved styrene and ethyl diazoacetate (EDA). However other alkenes (diphenyl ethylene ...) and other diazo compounds (*ie tert*-butyl and ethyl 2-phenyl diazoacetate) have also been investigated (Scheme 36).

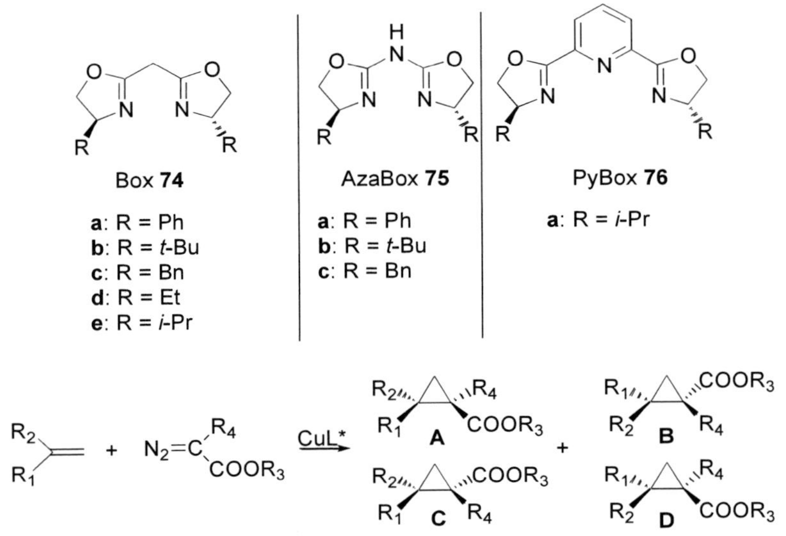

Box **74**

a: R = Ph
b: R = *t*-Bu
c: R = Bn
d: R = Et
e: R = *i*-Pr

AzaBox **75**

a: R = Ph
b: R = *t*-Bu
c: R = Bn

PyBox **76**

a: R = *i*-Pr

Scheme 36.

First, heterogeneous catalytic systems involved *N*-anchoring ligands on inorganic supports such as zeolithes [76], clays [77] ....

Immobilization of the catalyst on organic support leading to insoluble or soluble polymers was made by grafting or copolymerization of the chiral ligand.

In all cases, the immobilization is performed *via* alkylation of the methylene bridge or of the nitrogen of the bridge (AzaBox, PyBox).

3.1.1. Insoluble Supported-Polymer Box a Ligands

In 2000, Mayoral [78] described the first immobilization of Box on insoluble polymer. For a same ligand, grafting and several methods of polymerization (Scheme 37) were reported in order to compare the catalytic efficiency of these supported ligands after their complexation with Cu. The study was based on immobilization of bis(oxazoline) **74a-c** (Scheme 36) and their corresponding Cu-complexes were tested in the cyclopropanation of styrene with EDA, a well known reaction in homogeneous catalysis. [79]

The Box methylene bridge of **74a**, [78, 80] **74b** [78, 80, 81, 82] or **74c** [78, 80] was first functionalized with two *p*-vinylbenzyl groups in order to perform homopolymerization [78, 80] (polymer **79**), or *co*polymerization with styrene [78, 80, 81] (polymer **78**), DVB [78, 80, 81, 82] (polymer **80α**) divinylbenzyl polyethylene glycol [80, 81, 82] (polymer **80β**) and divinylbenzyl resorcinol [80, 81] (polymer **80γ**) or dendrimers **80δ** (Scheme 2). The best results were observed with the homopolymer **79b** (R = *t*-Bu) [78, 80]. This homopolymer **79** prepared by thermal polymerization [78] was able to coordinate a large amount of Cu and presented a better activity than the same homopolymer prepared in the presence of AIBN (respectively 51% and 34% yield, 35/65 and 39/61 *trans/cis* ratio, 75% ee and 77% ee for the *trans* cyclopropane, 70% ee and 73% ee for the *cis* cyclopropane) [78].

For all the copolymers, the low efficiency of the copper complexes (cyclopropanation of styrene: yield up to 40%) may be explained by the non accessible core of polymer by Cu probably due to the high degree of crosslinking. [78, 80] But moderate to high ee were observed (up to 78% ee) and contrary to homopolymers, the *cis* cyclopropane is mainly formed. Attempts to use copolymers from monofunctionalized Box (Scheme 38) as ligand did not improve the efficiency (yield up to 16%); the diastereo-selectivity is in favour of the *trans* stereoisomer. As for copolymer **80α**, yields obtained with **80β-δ**, do not exceed 35% and the diastereoselection gave mainly the *trans* cyclopropane. With **80β** the formation of *trans* isomer is up to 67%.[81, 83] The presence of oxygen atom in the crosslinking agents could induced more coordination with the metal. This effect is detrimental for the catalytic properties for the resulting Cu-loaded resin.[81]

The homopolymer **79b** could be reused up to five times without loss of ee but with a slight decrease in activity [80]

In order to avoid the non accessibility to the core of polymer by the metal due to excessive crosslinking, a monofunctionalization of the Box leading to Box pendant ligand **81** was performed [80] (Scheme 38). Unfortunately, the incorporation of copper into polymers **81** was not found to be improved (Cu loading: 0.07-0.08 mmol/g). Furthermore, a higher loading was observed with the less crosslinked polymer **81a**. It is noteworthy that if the ligand **81c** is transformed into a polymer with bis(oxazoline) moieties by treatment with

MeLi and MeI, the results obtained for the cyclopopanation reaction were slightly worse that those obtained with **81c** itself (yields 16% *vs* 16%; *trans/cis* ratio 60/40 *vs* 57/43, ee *trans* 23% *vs* 29%; ee *cis* 22% *vs* 34%).

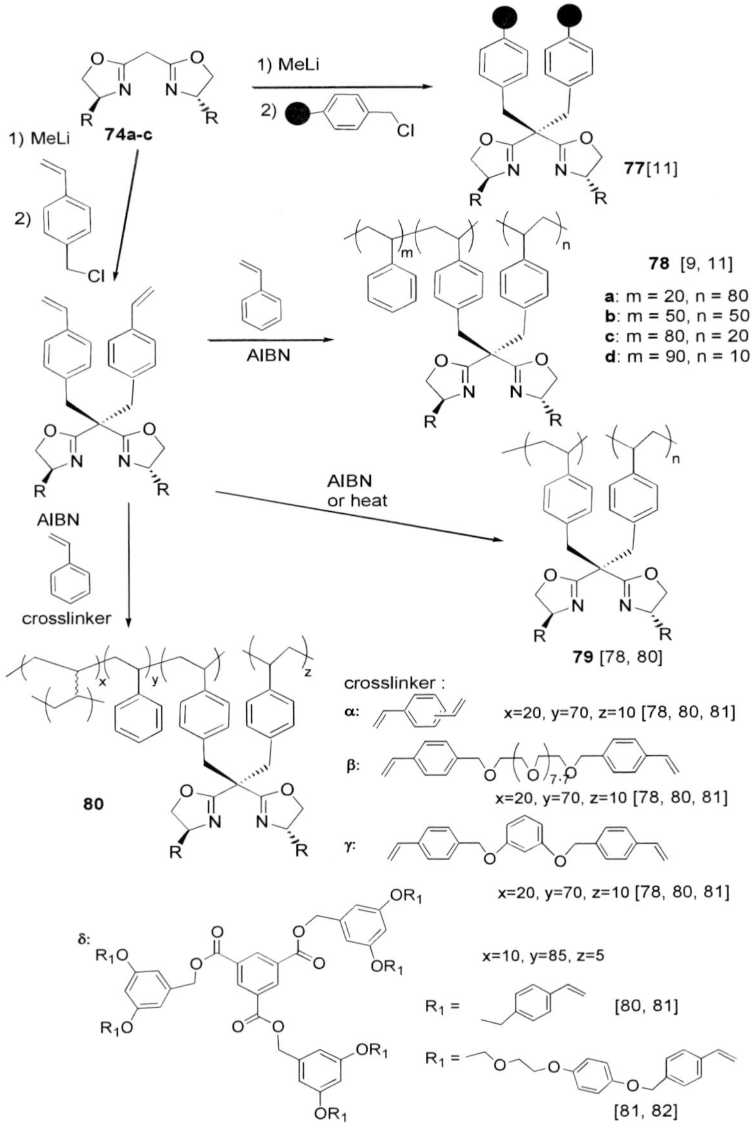

Scheme 37.

Salvadori [84] showed that a copolymer structure presenting a single spacer, but flexible one linking the ligand moiety to the inert polystyrene backbone was effective with solution-like behaviour (Scheme 39).

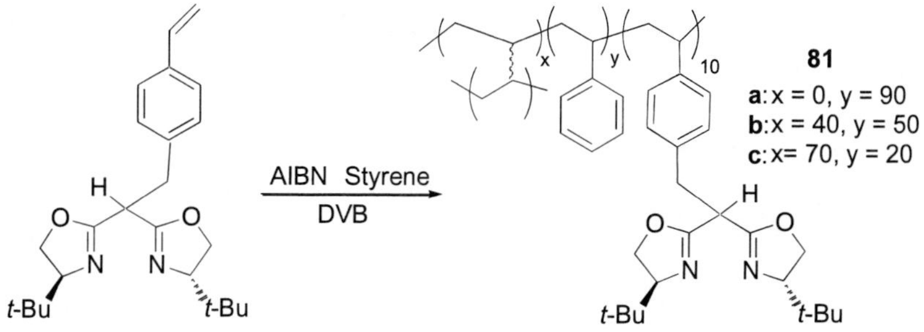

Scheme 38.

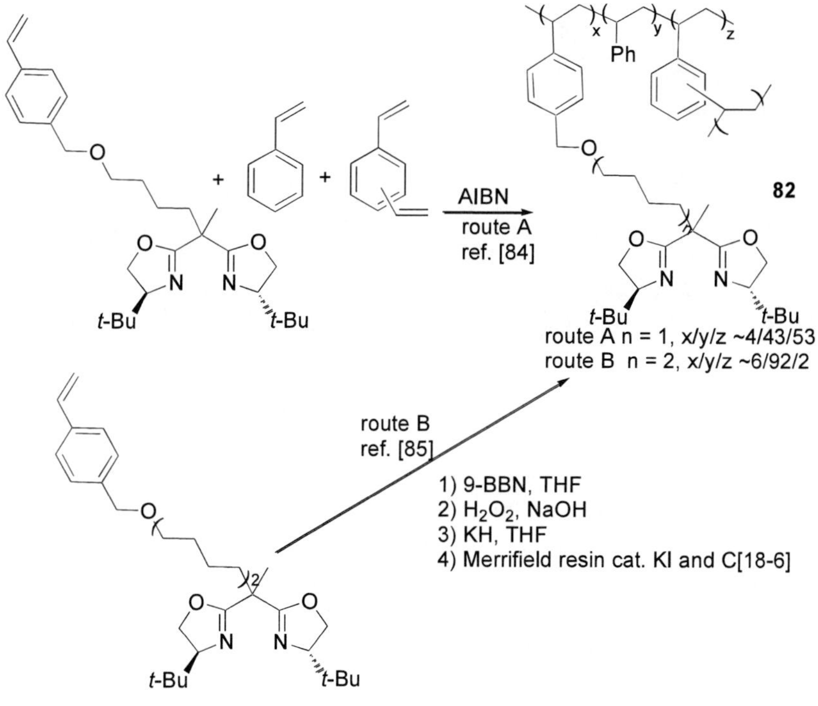

Scheme 39.

For the cyclopropanation of styrene with EDA, Cu-polymerized **82** complexes afforded similar results than the non supported Box. In the first case 61% and 60% yield, 67/33 and 71/29 dr in favour of the *trans*, 93% and 94% ee were respectively observed. The catalytic system could be reused without loss of its catalytic properties. Moreover, it is noteworthy that the immobilization method did not influence the catalytic behaviour: catalytic systems formed with copolymerized (route A) [84] or graft methodology (route B) [85] were substantially equivalent (variation of about 1 to 2% of the yield, ee or dr).

In 2004, Yifei reported Merrifield (3.8 mmol/g) polymer-supported Box **74c-e** [85] (Schemes 36 and 40) for the cyclopropanation of 1,1-diphenylethene with EDA.

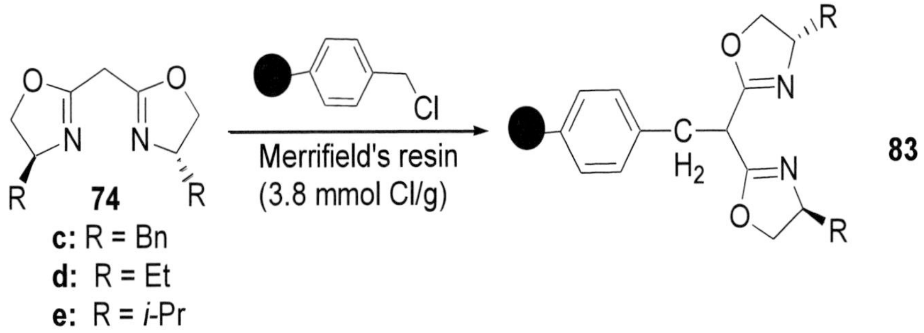

Scheme 40.

The highest enantioselectivity of 85% ee with 82% yield was achieved by using **83d** showing as in homogeneous conditions that the steric hindrance of the R group of the oxazoline had a great influence on the ee and the yield. Under recycling, the catalytic system led to an important decrease in enantio-selectivity.

In order to restrict the conformational mobility of the groups on the bridge of the Box, a 1,3-dioxane group was introduced, then the Box was grafted onto a Wang resin (0.7 mmol Br/g). (Scheme 41) [86]. In that condition, only 0.26 mmol/g has been grafted onto the active site and the copper content was 0.018 mmol/g. The ligand **84** was evaluated in the cyclopropanation of styrene with EDA. The *trans* cyclopropane was obtained in 65% ee. The high ee was due to less steric hindrance between the two oxazoline groups, as the consequence of the chair conformation of the dioxane ring.

AzaBox as well as PyBox have the advantage towards Box to present a donor nitrogen atom in the bridge allowing to a higher coordinating ability. Thus a better stabilization and a better recoverability, with regard to their analogous Box ligand are observed. Additionally, AzaBox and PyBox ligands have only one link point in the bridge which makes grafting easier.

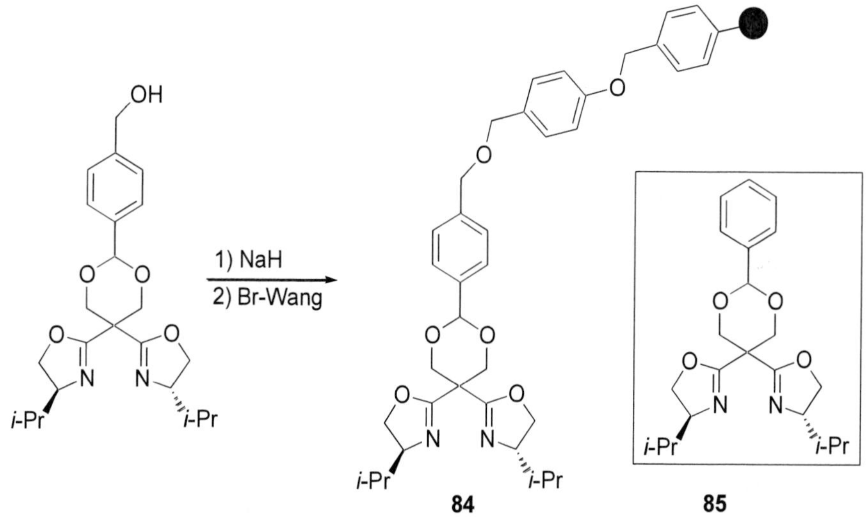

Scheme 41.

3.1.2. Insoluble Supported-Polymer AzaBOX as Ligands

Recently, AzaBox ligands were attached to a tentagel (polymer with a polystyrene backbone and PEG periphery) (**87**) or to polystyrene (**88** and **89**) by direct grafting of the ligand onto the polymer or by copolymerization of the ligand functionalized with a styryl group (Scheme 42) [87]. The corresponding catalyst presented a copper content of 0.2-0.3 mmol/g. Cyclopropanation of styrene and 1,1-diphenylstyrene with EDA was studied with 1-1.5 mol% of the catalyst. In these experiments, catalyst formed with copolymerized ligand **89** did not show any advantage over grafting ones. Moreover, Merrifield resin was a better support than tantagel. With **87a** the yield was up to 51% but its diastereoselection (64/34) and and its ee (up to 62%) are lower than those obtained with **88** or **89** (yield up to 32%, dr 70/30, ee up to 88%) for the cyclopropanation of styrene.

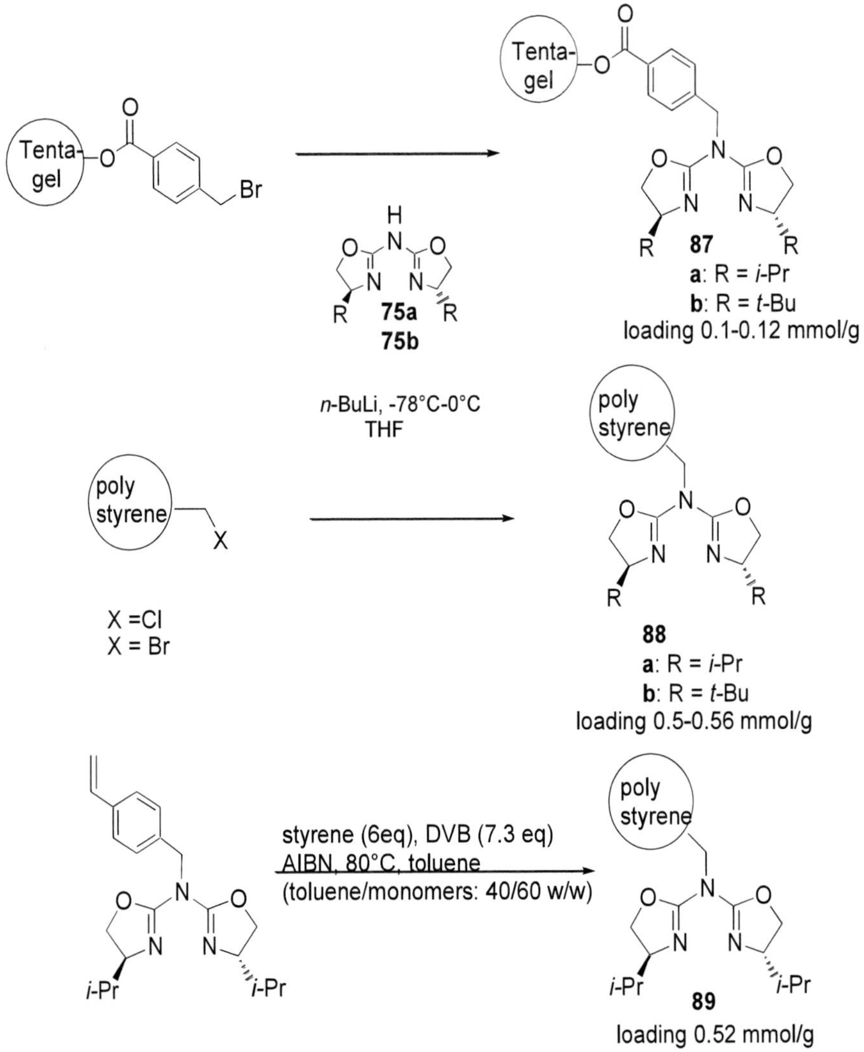

Scheme 42.

3.1.3. Insoluble Supported-Polymer PyBox as Ligands

The first immobilization of PyBox was made by Mayoral [88] by means of radical copolymerization of 4-vinylPyBox with styrene and DVB in the presence of AIBN and a porogen (toluene or a mixture toluene/dodecanol) (ligand **90** Scheme 43). PyBox was incorporated in the polymer with high yield from 75 to 95%. The corresponding Ru-complex was formed at about 50-60% showing that all the ligands were not accessible. The nature of the porogen solvent was important. A polymeric ligand prepared in the presence of dodecanol gave a lower Ru functionalization and was detrimental to the *trans/cis* selectivity (77/23). The best catalytic system, high *trans/cis* selectivity (85/15) and ee (*trans*: 85%), was formed by the polymer prepared in toluene. However, it has been observed that with a lower crosslinking degree, dodecanol has less importance. The best catalysts were reused twice, but a marked decrease in both selectivities and activities were observed for the second recycle.

Subsequent studies involving not only copolymerized ligands but also polymers grafted onto a Merrifield resin (Scheme 43) were performed [89]. For the latter methodology, functionalization of the PyBox was required, thus 4-aminophenoxyPyBox and 4-mercaptoPyBox were synthesized, leading to the corresponding ligands supported polymers **91-93**. For the Ru catalysts, the introduction of a spacer is important. The longer the spacer was, the higher Ru functionalization. This is consistent with the better accessibility of the grafted ligands contrary to copolymerized ones. Concerning their evaluation for the cyclopropanation of styrene with EDA it has been found that to obtain high catalytic performance and good recyclability, the heterogeneous catalyst must have a gel type organic support with a low crosslinking (2% DVB) and a support with no electron-donating effect on the pyridine rind (C-C bond between the support and the ligand). Thus taking into account these para-meters, a highly active (40-65% yield) and enantioselective (87-91% ee) catalyst was obtained. Reuse at least 3 times of the catalytic system with the same performance was possible. Mayoral [90] also demonstrated that the PyBox-Ru supported complexes on macroporous monolithic polymers were efficient catalysts for the cyclopropanation reaction. PyBox monolithic minireactors were prepared in a stainless column by radical copolymerization of 4-vinyl or 4-styryl PyBox in the presence of styrene and DVB using 10% toluene/50% dodecanol as the precipitating porogeneous mixture (polymers **94** and **95** Scheme 44). To form the Ru complex, a solution of [RuCl$_2$(*p*-cymene)]$_2$ was passed through the column at low flow then washed with dichloromethane to remove the non complexed Ru.

For the cyclopropanation of styrene with EDA, several parameters have been studied: flow rate, morphology of the catalyst and solvent.

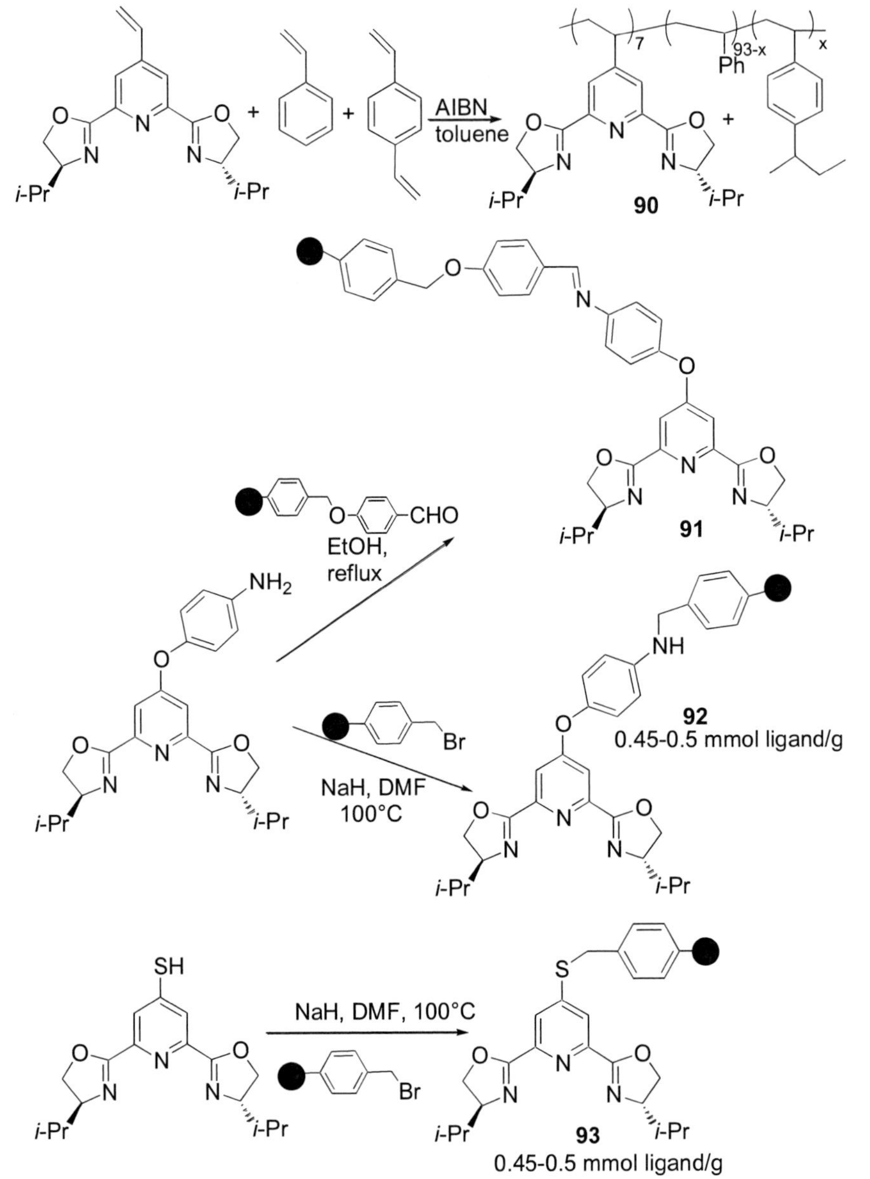

Scheme 43.

An increase of the flow rate induced a decrease of the cyclopropanation and a dimerization of EDA which is one of the poisons of the catalyst. Besides both region and enantioselectivities gave good values: 80/20 *trans/cis*, *cis* isomer 48 ± 5% ee, *trans* isomer 75 ± 5% ee. The catalyst could be used for several runs (4-6 times) without change in catalytic performance. Concerning the morphology of the catalyst, a slight decrease in catalytic activity was found with ligand **95** containing an aryl spacer. If the crosslinking degree was higher (51% DVB instead of 20%), the activity and chemoselectivity were improved. The reaction performed without solvent provided an increase in both efficiency and chemoselectivity for a mixture styrene/EDA: 1/7 and minimize the dimerization of EDA. With scCO$_2$ as solvent, yields were about 25% but chemoselectivities and ee of both diastereoisomers rise. Moreover, the total number of turnover (TON) is greatly increased.

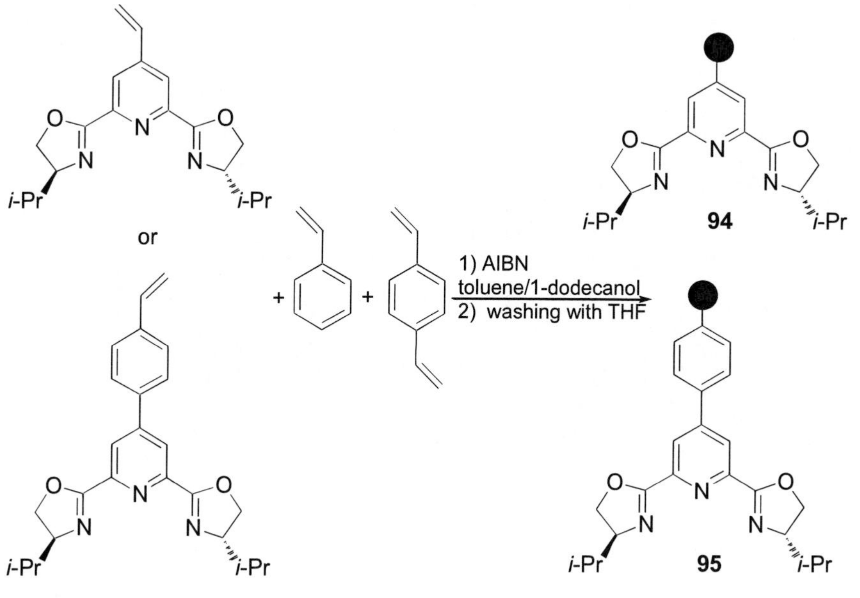

Scheme 44.

3.1.4. Immobilization of PyBox on Natural Polymers

Some natural polymers have also been used as support for Schiff base ligands. A starch support has also been used as support for a PyBox. A telomerized starch **96** [91] presenting a final double bound reacted with a thiol-PyBox in the presence of AIBN (Scheme 45) to form the corresponding immobilized PyBox. [92] Although the Ru-**97** complex exhibited lower activity (up to 44% after the third reuse) and selectivity (*trans/cis*: 76/24 with

50% ee for the *trans* and 16% ee for the *cis*) towards the homogeneous complex formed with **98** (yield 67%, *trans/cis*: 89/11 with 77% ee for the *trans* and 63% ee for the *cis*) this supported catalyst showed better performance than that immobilized on silica [93]. These results pointed out a great influence of the support.

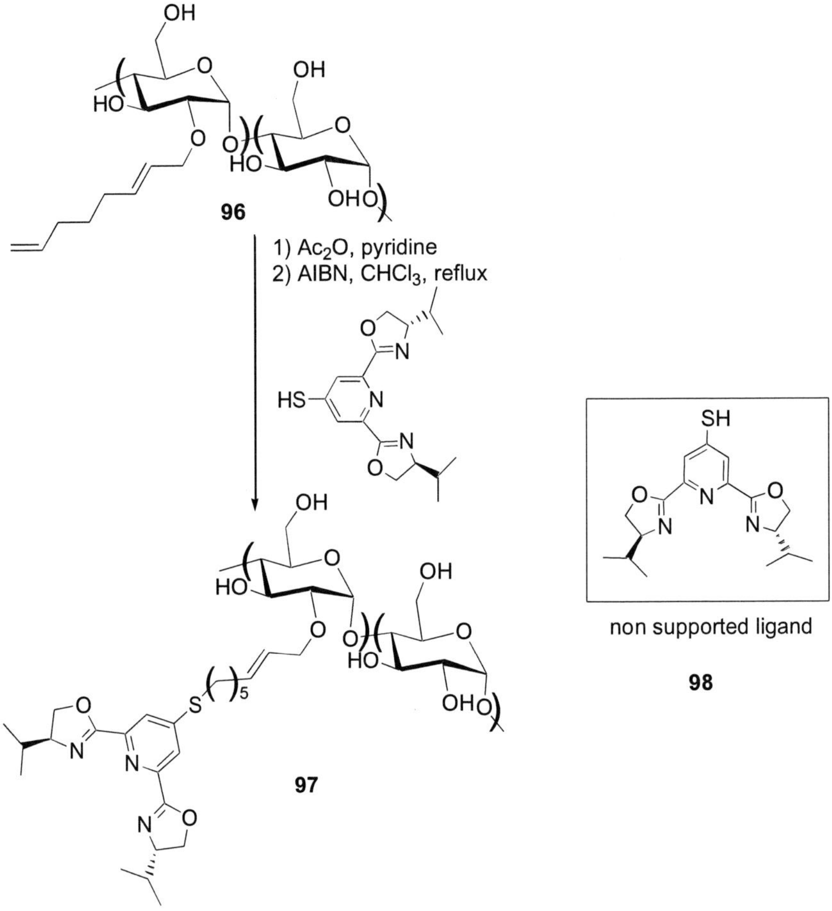

Scheme 45.

The homogeneous catalyst formed by reaction of non supported oxazoline **98** with RuCl$_2$ was very efficient, the ratio *cis/trans* was about 89/11 and enantioselectivities were 63% and 71% respectively. These values are lower than those achieved using 4H PyBox-Ru catalyst (88 and 70% respectively), owing to the presence of a slightly electron-withdrawing group on the pyridine moiety of the ligand. The Ru-complex of the polymer supported ligand **97** exhibited lower activity and selectivity toward the cyclopropanation reaction of styrene. Three recycles were performed showing a *trans/cis* ratio and the enantioselectivity constants.

3.1.5. Soluble Supported-Polymer Box as Ligands

Box **74a** and **74b** have also been immobilized onto the soluble polymer MeOPEG (a monomethylether of polyethylene glycol, Mn > 2000 Da) and employed as ligand in the Cu-catalyzed cyclopropanation of styrene and 1,1-diphenylethylene with EDA with up to 93% ee.

Taking into account the best parameters found with insoluble polymer, which were necessary to enhance the stereoselectivity of Box promoted reaction [94], *ie* disubstitution at the Box bridging atom and the necessity of introducing a spacer in order to separate the polymer moiety (PEG) to the ligand itself, Cozzi synthesized the supported polymers **99a** and **99b** (Scheme 46) [95]. With styrene, ligand **99b** gave a *trans/cis* stereoselectivity in favour of the *trans* product (70%), the *trans* isomer presenting 87% ee. An increase of the yield (about 20%) was observed if the time of addition of EDA and of the reaction increased. In the same conditions, the results were inferior to those obtained with the *gem* dimethyl Box **74b** but the two MeOPEG-supported catalysts presented similar behaviour. Recycling of the catalytic system was possible with marginal erosion of the catalytic activity and very limited loss in ee.

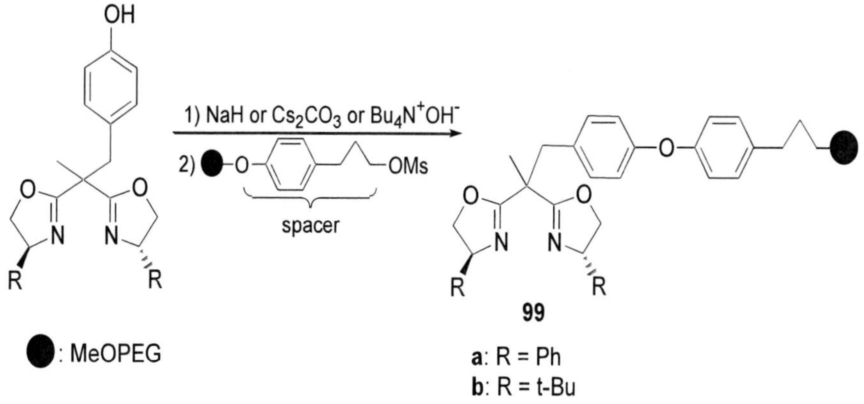

Scheme 46.

3.1.6. Soluble Supported-Polymer AzaBox as Ligands

In 2000, Reiser [96] reported the first immobilization of aza-bis(oxazolines). Aza-bis(oxazoline) **74b** has been easily grafted to a soluble polymeric support ((methoxy(polyethyleneglycol) MeOPEG 5000) in order to form soluble catalysts which could be recovered by precipitation. This polymer (**100** Scheme 47) was obtained successfully (55% yield) if a benzylidene spacer linked to the PEG was employed. Complexed with Cu(II), this ligand was able to promote asymmetric cyclopropanation of styrene (Scheme 36: R_1 = Ph, R_2 = H, R_3 = Me, R_4 = H) and 1,1-diphenylethene (scheme 36: R_1 = R_2 = Ph, R_3 = Me, R_4 = H) with methyl diazoacetate in dichloromethane. With styrene, a diastereoselectivity in favour of the *trans* cyclopropane with a predominance of **C** over **D** was observed (**C**/**B**: 71/29, **C**: 91% ee **B**: 87% ee) with a yield of 69% (Scheme 36). Compared to the non-supported catalytic system (ligand **75b**), the results in terms of diastereoselection and ee were 10% lower. The possibility of the reuse of this catalytic system was examined. 13 cycles were conducted without loss of ee, but to keep a good yield after the 10th use, it was necessary to reactivate the catalytic system with phenyl hydrazine.

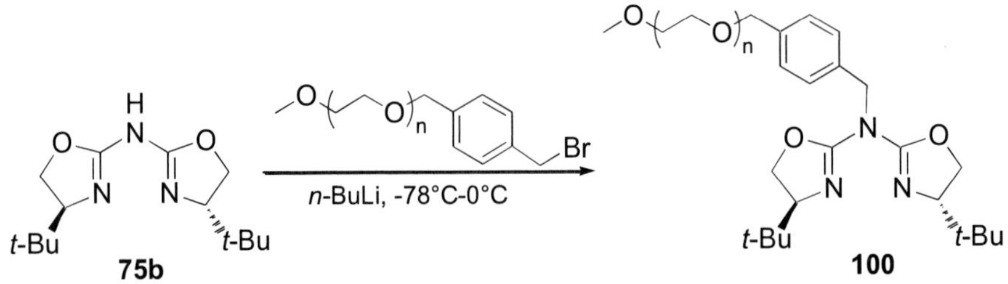

Scheme 47.

3.2. PORPHYRINS

More recently, chiral Ru-porphyrin polymers were used for catalytic asymmetric carbene transfer. A C2-symmetric group containing two norbornane moieties fused to the central vinyl substituted benzene ring of a porphyrin was chosen in order to induce the chirality. Then the chiral ruthenium vinylporphyrin **101** was involved into radical copolymerization with styrene and DVB or EGDMA to lead respectively to monolithic resins chiral **102** and **103** which were crushed (Scheme 48).

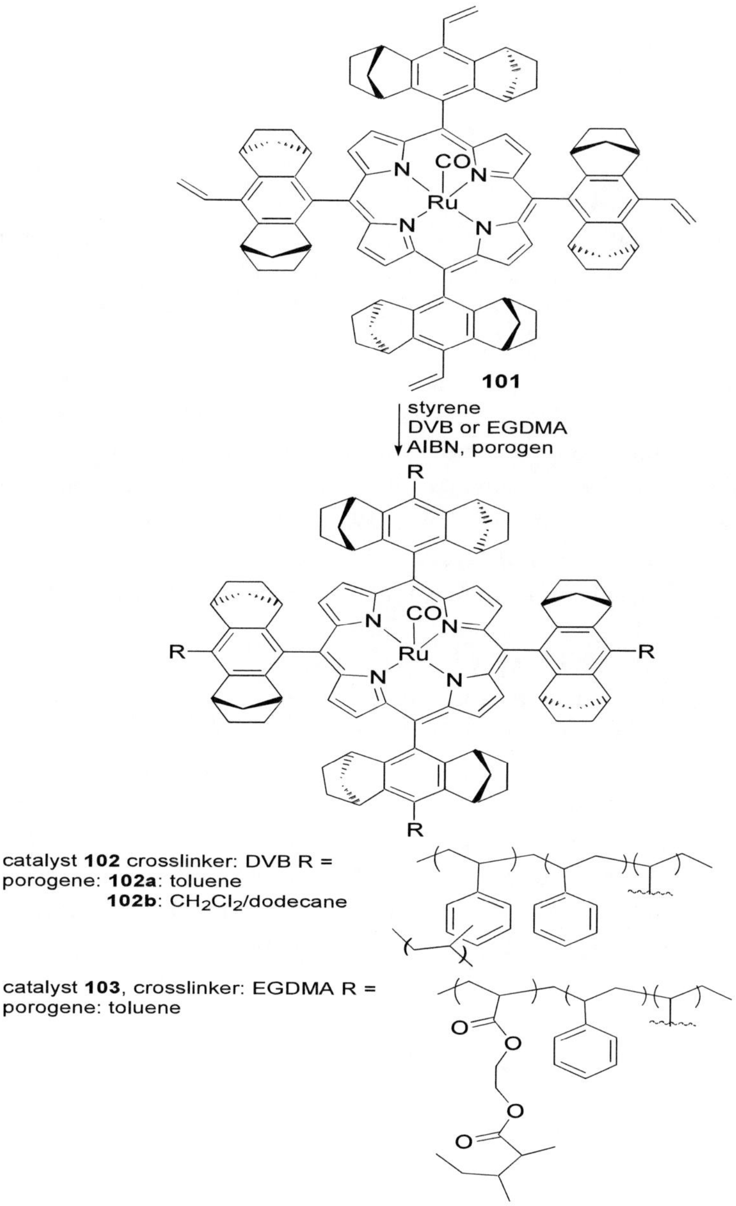

catalyst **102** crosslinker: DVB R =
porogene: **102a**: toluene
         102b: CH$_2$Cl$_2$/dodecane

catalyst **103**, crosslinker: EGDMA R =
porogene: toluene

Scheme 48.

In order to form cyclopropane esters and cyanocyclopropanes, EDA and diazoacetonitrile were respectively used in the cyclopropanation of various substituted styrenes. Although less enantioselective than their homogeneous counterpart, these catalysts in the presence of EDA led to high stereo-selectivities. For example with **102a**, up to 92/8 for styrene and *p*-methoxystyrene and 97/3 for *p*-bromostyrene, in favour of the *trans* isomer were obtained. Yields were respectively 77%, 88% and 75%. The best ee (90%) was obtained with the *p*-bromo derivative. Moreover, diastereo-selectivity and enantioselectivity were influenced by the nature of the porogenic solvent employed during the formation of the catalyst. Compared to **102b**, better ee and diastereoselectivities were obtained with catalyst **102a** (*trans/cis* ratio: 82/18 and 92/8; 82% ee and 71% ee respectively), these two catalysts having the same DVB/styrene ratio. This behaviour was attributed to the reduced accessibility of the catalytic sites, also evidenced with polymers prepared from higher DVB/styrene ratio. With 102 slightly lower results than those of **102a** were observed. Concerning the cyclopropanation with diazonitrile, moderate activities from 35 to 53% yield and correct enantioselectivities around 70% were obtained, the best results being found with catalyst **102a**. These catalysts were stable upon recovery and recyclability (no leach of metalloporphyrins) but a sligh decrease in activity was observed [97].

Supported metalloporphyrins have also been synthesized by electro-polymerization leading to films of polymers. They were formed on the platinium working electrode during the oxidative electrosynthesis (Scheme 49) [98]. Contrary to the preceding catalysts **102** and **103**, these electro-polymerized catalysts led to low enantioselectivities for the bench-mark reaction between styrene and EDA. Up to 53% ee at -40°C was reached in the presence of **105a**. Moreover, at room temperature, the reactions proceeded efficiently (yields 80-90%). Seven recycling of polymer **105a** were carried out without a significant decrease in enantioselectivity and activity.

3.3. MISCELLANEOUS

Instead of Box ligand, carboxamidate coordinate with rhodium have been tested successfully for the cyclopropanation of some functionalized alkenes [99] (Scheme 50).The carboxamidate anchored to the NovaSyn Tentagel hydroxyl resin (TG) or Merrifield resin (Scheme 50) was treated with chiral

dirhodium   tetrakis[methyl   2-oxypyrrolidine-5(*S*)-carboxylate]   (Rh₂((*S*,*S*)
MEPY)₄) leading to the corresponding Rh-**106** and Rh-**107** precatalyst.

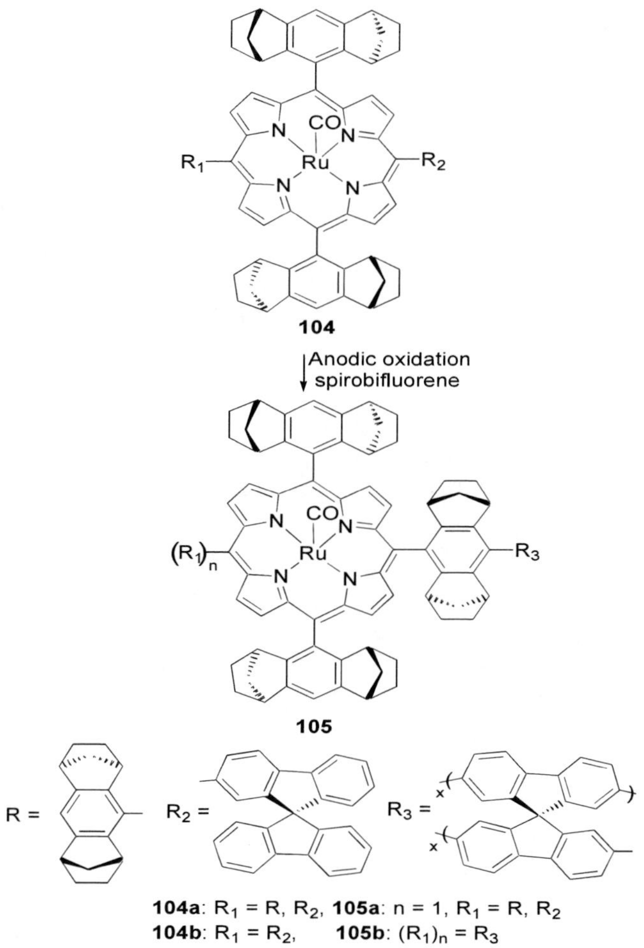

104

Anodic oxidation
spirobifluorene

105

R =   R₂ =   R₃ =

104a: R₁ = R, R₂, **105a**: n = 1, R₁ = R, R₂
104b: R₁ = R₂,   **105b**: (R₁)ₙ = R₃

Scheme 49.

With these catalysts, intramolecular cyclopropanation of allyldiazoacetate
was investigated (Scheme 51). Compared with homogeneous catalyst, yield of
bicyclolactone was similar to those obtained with the two catalysts *ie* 75% and
95% ee.

Catalyst with the Merrifield support (Rh-**107**) presented a yield of about
75% and a slight decrease in ee after 10 runs contrary to the TG one (Rh-**106**)
which showed 10% of the yield and of about 20% ee decrease. Ligand loading

had an influence only on the yield but not on the selectivity of the reaction. A similar study was performed with the cyclopropanation of styrene and EDA. Once again, the best catalytic system in terms of reproducibility of the results on the reuse corresponds to the Merrified one. For the first use they were quite equivalents.

In 2003, Xia [100] has reported the cyclopropanation of styrene with alkyl diazoacetates by copper complexes of Shiff bases, derived from chitosan and substituted salicylaldehydes as the catalyst (Scheme 52). This natural polymer is an environmental friendly natural material: biocompatible, biodegradable, non toxic...

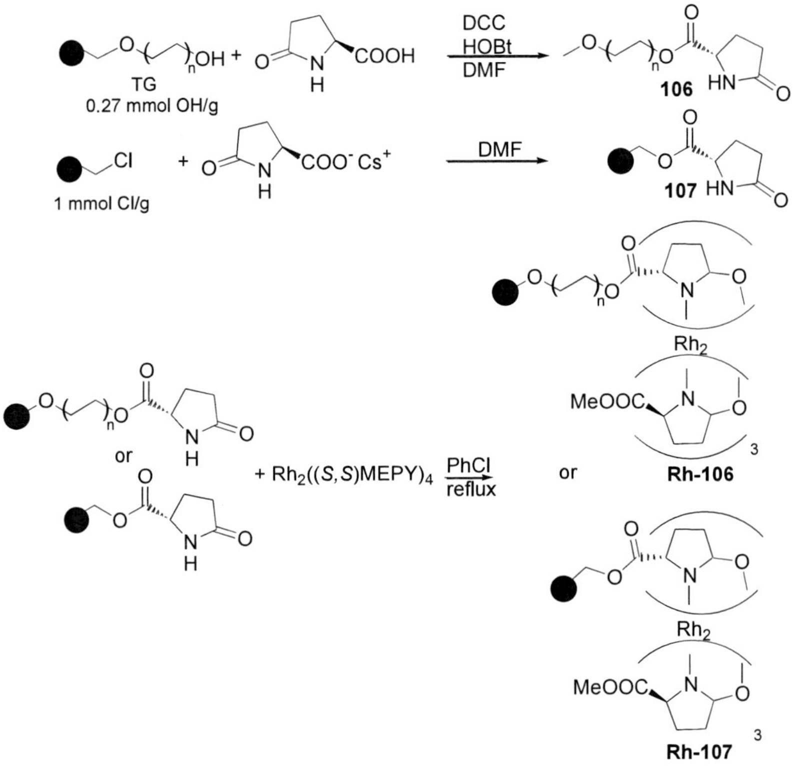

Scheme 50.

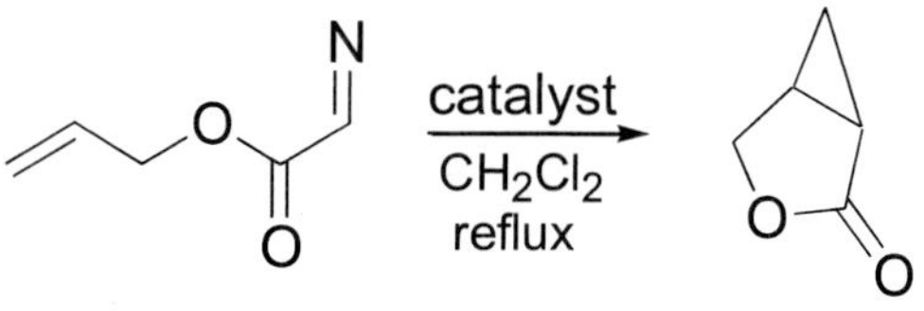

Scheme 51.

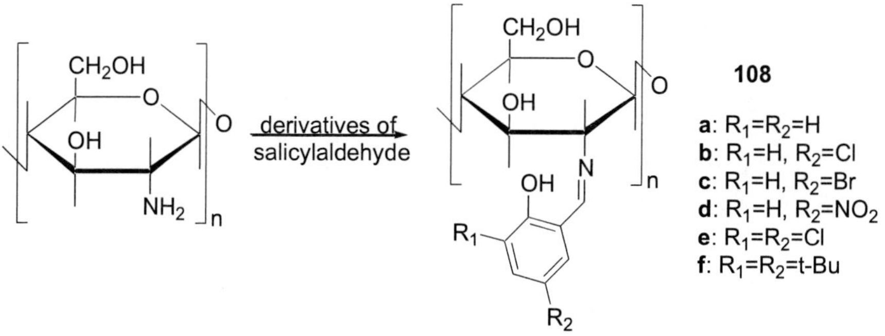

108

a: $R_1=R_2=H$
b: $R_1=H, R_2=Cl$
c: $R_1=H, R_2=Br$
d: $R_1=H, R_2=NO_2$
e: $R_1=R_2=Cl$
f: $R_1=R_2=t\text{-Bu}$

Scheme 52.

The influence of several parameters: nature of the solvent, temperature, the molar ratio styrene/EDA and the copper content [100] have been studied. Better yields were obtained with high temperature but better ee were found at 60°C, temperature for which the nature of the solvent had a great influence. From all the solvent tested, 1,2-dichloroethane, ethylacetate, acetonitrile and toluene, the best yield (91.5%) was obtained with 1,2-dichloroethane; toluene led only to 41.5% yield. Better ee were also found in 1,2-dichloroethane. When the molar ratio styrene/EDA was about 17/1 the results of this reaction were fairly good; as this ratio diminished, the yield and ee decreased and for a ratio of 2/1, the yield reached only 45.3%. Taking into account of all these parameters, it has been shown that to obtain the best compromise between the yield (81.5%), the best *cis/trans* ratio (35.3/64.7) and the best ee (*cis*: 20.7% ee, *trans*: 10%ee) for the cyclopropanation of styrene, a copper content (Cu(OAc)$_2$) of 4.4 mol% was necessary with ligand **108a**. More or less amount of the copper content led to a decrease of all these reaction results. It is noteworthy that neither the electron-withdrawing nor the bulky substituents on the salicylaldehyde of the catalysts had much influence on the yield, the ee of the *cis* and *trans* isomers but the nature of the diazoacetate had a great influence. Using *t*-Bu diazoacetate instead of the Et one resulted in a important increase of the ee for both the *cis* and *trans* isomers but with a dramatic drop of the yield. This behaviour could be a result of the steric hindrance of the *t*-Bu groups which provoked a difficulty to form metal carbene species by the reaction of *t*-Bu diazoacetate with the Schiff base copper complexes. These catalytic systems have also been use to transform other alkenes into cyclopropyl derivatives. In some cases, compared to the styrene reaction the cyclopropanation of 1-hepten or 1-octene gave better ee.

To conclude, for the cyclopropanation reaction, the difficulties consist in the achievement of high enantioselectivity close to a good stereoselectivity. To be useful in organic synthesis, the last requirement is preponderant and depends on the nature and the number of the substituents of the cyclopropane ring. It is the reason why, most of the catalytic systems used in heterogeneous supported catalysis cyclopropanation are formed by bis(oxazoline)s complexes which have shown a excellent efficiency in non supported conditions.

In most of the cases, this reaction seems to be very sensitive to the polymer morphology controlled by crosslinking and the nature of the porogen agent which determine the performance of the immobilized catalyst.

Chapter 4

CYCLOADDITIONS
AND HETEROCYCLOADDITIONS

Discovered in 1928, the Diels-Alder reaction is one of the powerful synthetic methods for the construction of substituted cyclohexene that fits the modern concept of atom economy [4]. For this reason, in the last two decades, the asymmetric Diels-Alder reaction, leading to carbon-carbon bond formation, has received considerable attention and enantioselectivities greater than 90% have been reported [101]. In homogeneous conditions, a wide range of metal, ligands and dienophiles have been studied. Because of their high stabilities, most of the studies are focused on the use of *N*-containing complexes.

Concerning the Diels-Alder reaction, polymeric supports were mostly involved in solid phase synthesis where the diene and the dienophile are grafted onto the polymer. Compared to these procedures, only few examples involving supported catalysts were reported in the literature.

Most of these supported catalysts contain nitrogen as ligands such as Box, aminoalcohol, *N*-sulfonylamino acid and salen.

4.1. SUPPORTED BOX LIGANDS

Moberg [102] developed a polymer bound bis(oxazoline) **109** (Scheme 53) as a zinc complex catalyst for the Diels-Alder reaction of cyclopentene with 3-(2-propenoyl)-2-oxazolidinone (Scheme 54). This supported catalyst was formed by grafting a phenolic bis(oxazoline) derivative onto an ArgoGel Wang-Cl resin and presented 0.071 mmol of ligand/g polymer.

Compared to the monomer (conversion 100%, *exo/endo* ratio: 94/6, ee: 85% for the *exo* adduct), the polymer supported catalyst gave no selectivity (racemic compounds are formed) and lower reactivity (13% at -40°C). This behaviour is probably due to the low swelling and the rigidity of the polymer at low temperature.

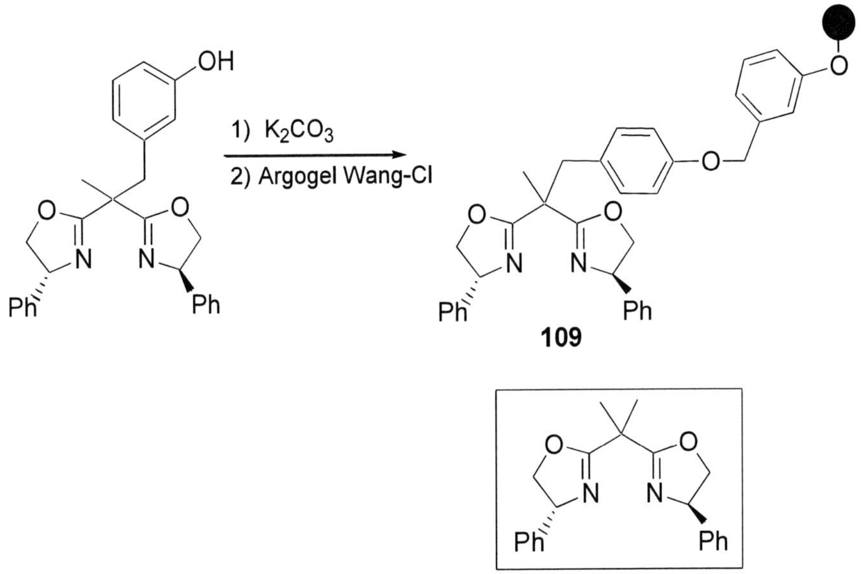

Scheme 53.

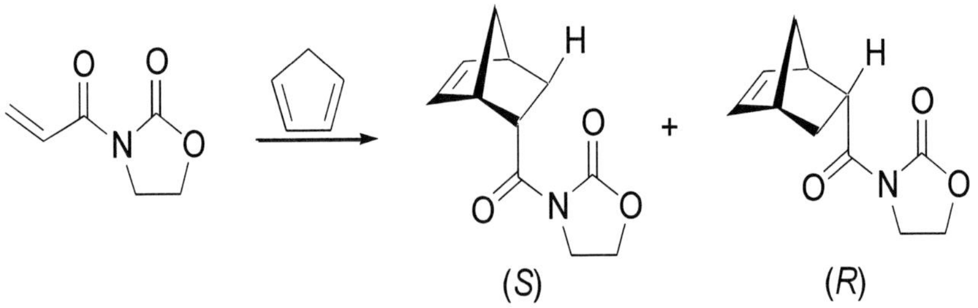

Scheme 54.

At the same time, Cozzi described the immobilization of Box ligands on PEG polymer (polymer **99**, Scheme 49), and the corresponding copper catalyst used for the cyclopropanation reaction was also involved in Diels–Alder cycloaddition between cyclopentadiene and *N*-acryloyloxazolidinone (Scheme 54) [95]. This reaction was performed with 15 mol% each of Box ligand and $Cu(OTf)_2$ in DMF in order to obtain homogeneous conditions. The reaction occurred in good yield (up to 83%) and excellent diastereoselectivity in favour of the *endo* stereoisomer (ratio *endo/exo* >98/2). However, the maximum of enantioselectivity was only 45% contrary to the non-supported Box equivalent catalyst which led to complete end at 95% ee and 88% yield.

In 2002, Lemaire chosed to immobilize indaBox, not only for its rigidity but also for its good enantioselectivities for the Diels-Alder cycloadditions [103]. The heterogeneization was carried out by polymerization as part of the main chain of a polymer backbone (Scheme 55). Mixed with $Cu(OTf)_2$, polymer **110** (8 mol% relative to the oxazolidinone) was allowed to react with acryloyl oxazolidinone and cyclopentadiene at -78°C (Scheme 55). The reaction was quantitative and led to the major *endo* diastereoisomer with 51-56% ee with 87-90% yields for the first three reaction cycles.

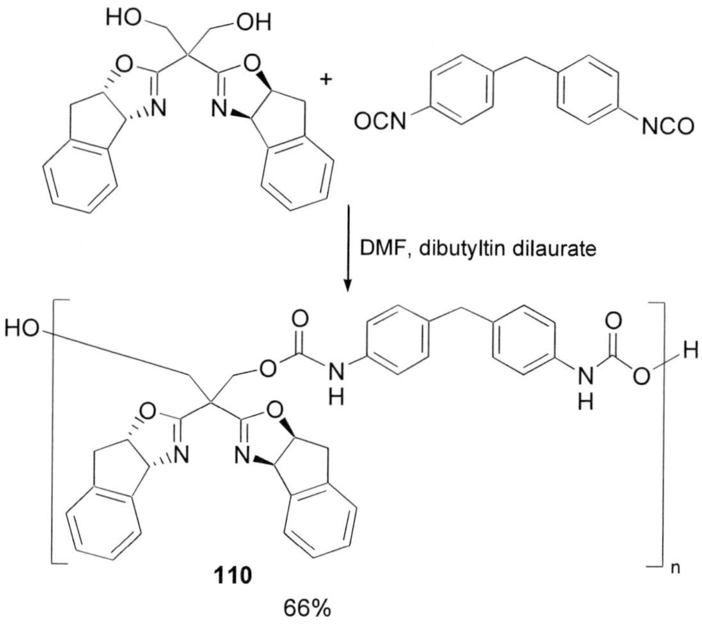

Scheme 55.

4.2. SUPPORTED AMINOALCOHOL AND *N*-SULFONYLAMINO ACID-DERIVED OXAZABOROLIDINE OR OXAZABOROLIDINONE

Itsuno synthesized supported aminoalcohols **111-113** (Scheme 56) [104] and *N*-sulfonylamino acid **114** [105-107] (Scheme 57) by radical suspension copolymerization of the corresponding monomer chiral styryl aminoalcohols and styryl sulfonylamino acid derivatives with styrene and eventually with a crosslinker.

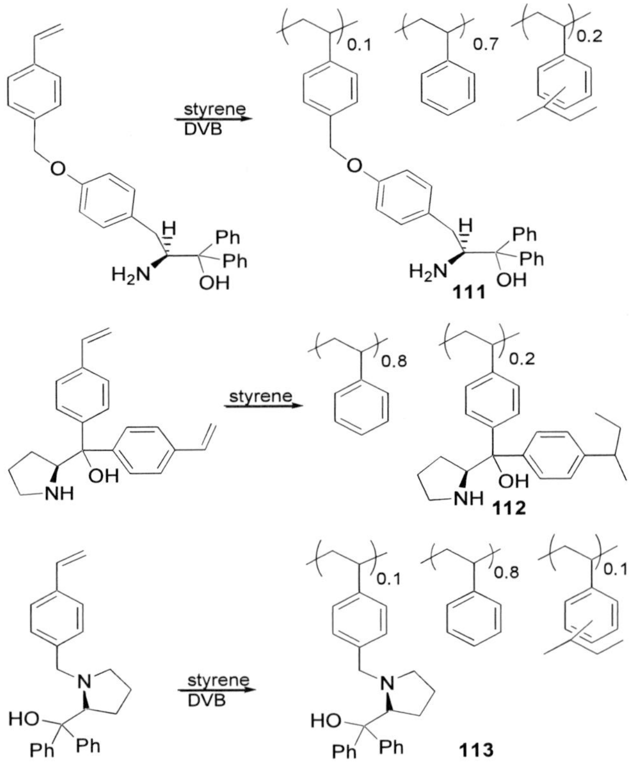

Scheme 56.

The styryl chiral sulfonylamino acid derived from L-valine **114a**, L-isoleucine **114b**, L-threonine **114c** or D-2-phenylglycin **114d** (Scheme 57). The loading of aminoacid residue was about 0.78-0.88 mmol/g for a polymer formed with 10% of the styryl chiral sulfonylamino acid, 80% styrene and

10% of DVB with a yield range of 85-96%. In the case of valine, for a polymer formed with 50% of the styryl chiral sulfonylamino acid, 40% styrene and 10% of DVB, it could be up to 2.55 mmol/g and 73% yield [105].

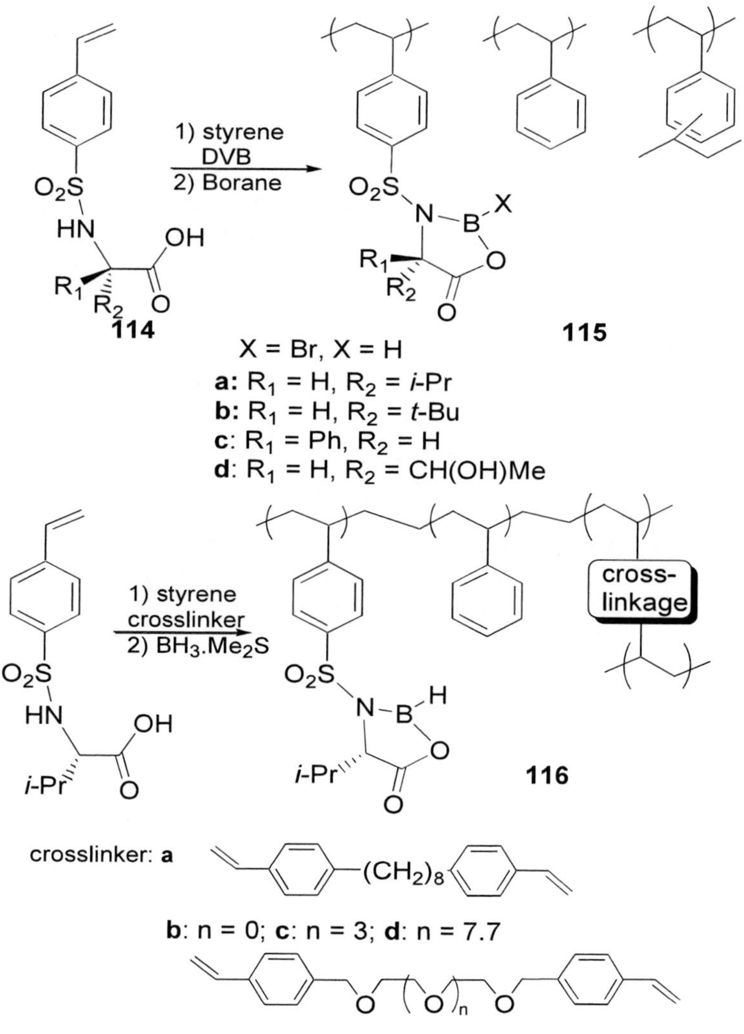

1) styrene
DVB
2) Borane
O2S
HN
OH
R1
R2
O
114
115
X
N
B
O
X = Br, X = H
a: R1 = H, R2 = i-Pr
b: R1 = H, R2 = t-Bu
c: R1 = Ph, R2 = H
d: R1 = H, R2 = CH(OH)Me
1) styrene
crosslinker
2) BH3.Me2S
i-Pr
cross-
linkage
H
116
crosslinker: a
(CH2)8
b: n = 0; c: n = 3; d: n = 7.7
n

Scheme 57.

The Diels-Alder reaction was performed with cyclopentadiene and methacroleine (Scheme 58) in the presence of 15 mol% of the oxazaborolidine and oxazaborolidinone catalysts derived respectively from supported aminoalcohols and from *N*-sulfonylamino acid polymers. The oxazaborolidine and oxazaborolidinone catalysts were formed *in situ* by action of BH_3, BH_2Br, $BHBr_2$ or BBr_3. The diastereoselectivity was excellent in favour of the *exo* adduct and yields from 65 to 99%. It is noteworthy that higher loading of chiral catalyst site in the polymer, lower *exo* selectivity and enantioselectivity. The diastereoselection depended not only on the nature of the supported ligand, the crosslinker but also on the borane and the solvent. Results are summarized in Table 6.

From all the oxazaborolidine catalyst, the L-valine derived catalyst **115a** gave the highest enantioselectivity and an excellent ratio *endo/exo*. Moreover oxazaborolidinone polymeric catalysts having oxyethylene crosslinkages (**116**) present better performance.

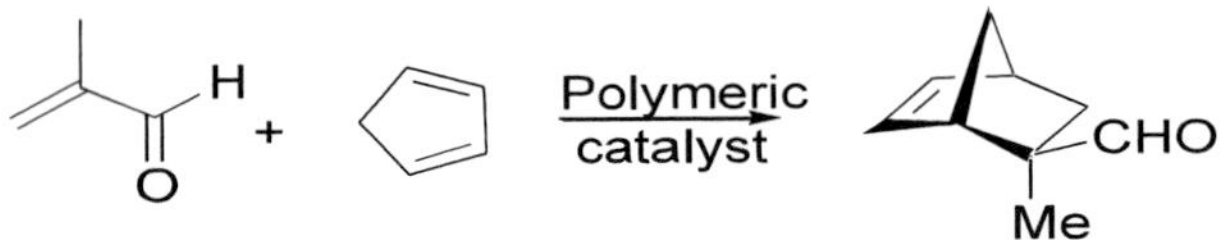

Scheme 58.

Table 6. Asymmetric cycloaddition reaction of cyclopentadiene and methacroleine in the presence of 15 mol% polymeric boron catalyst, at -78°C

Polymer	Borane	Solvent	Yield (%)	*Endo/exo*	Ee (%)	Ref.
111	BH_2Br	CH_2Cl_2	96	8/92	16	[104]
112	BH_2Br	CH_2Cl_2	98	4/96	25	[104]
113	BH_2Br	CH_2Cl_2	99	< 1/99	44	[105]
113	BH_3	CH_2Cl_2	95	< 1/99	57	[104]
113	BH_3	CH_2Cl_2/THF	89	< 1/99	65	[104]
113	BBr_3	CH_2Cl_2/THF	96	5/95	54	[105]
115a	BH_3	CH_2Cl_2/THF	87	< 1/99	65	[106] [107]
115b	BH_3	CH_2Cl_2/THF	99	3/97	49	[106]
115c	BH_3	CH_2Cl_2/THF	99	4/96	49	[106]
115d	BH_3	CH_2Cl_2/THF	65	8/92	10	[106] [107]
116a	BH_3	CH_2Cl_2/THF	86	8/92	84	[107]
116b	BH_3	CH_2Cl_2/THF	85	5/95	77	[107]
116c	BH_3	CH_2Cl_2/THF	93	1/99	92	[107]
116d	BH_3	CH_2Cl_2/THF	88	4/96	95	[107]

4.3. SUPPORTED SALEN AS LIGAND

Seebach [108] prepared non-dendritic **117** and dendritic **118** and **119** styryl-substituted salen, monomers for crosslinking suspension copolymerization with styrene (Schemes 59 and 60). Complexed with Cr, these resulting copolymers were involved in Diels-Alder reaction of Danishefsky's diene with several aldehydes (Scheme 61). In all cases, these new Cr-salen catalysts led to slightly lower enantioselectivities than those observed with their homogeneous analogs under the same conditions. Surprisingly, upon reuse, enantio-selectivities generally increased around 5% from the first to the fifth catalytic cycle. With caproaldehyde and with benzaldehyde enantioselectivities are respectively higher than 85% and 75%.

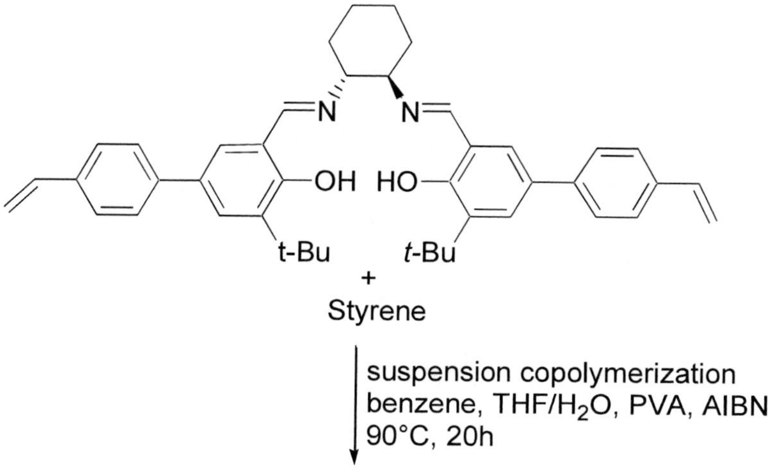

Scheme 59.

Recently, electropolymerization, an original methodology to prepare polymer supported salen was carried out by Schulz **113** [109][110]. It consisted in the introduction of a thiophene moiety, an electropolymerizable functionality, on the salen backbone (Scheme 62) followed by electro-polymerization under cyclic voltammetry conditions on a platinum grid. These polymeric catalyst **120a** and **120b** were then evaluated in hetero Diels-Alder reaction (HDA) between several aldehydes and 1-methoxy-3-[(methylsilyl) oxy]-1,3-butadiene (Scheme 61) using 4 mol% **120a** or 6 mol% **120b**. It has been shown that the counter ion has quite no influence on the selectivity but the activity is lower with **120b** containing BF$_4$.

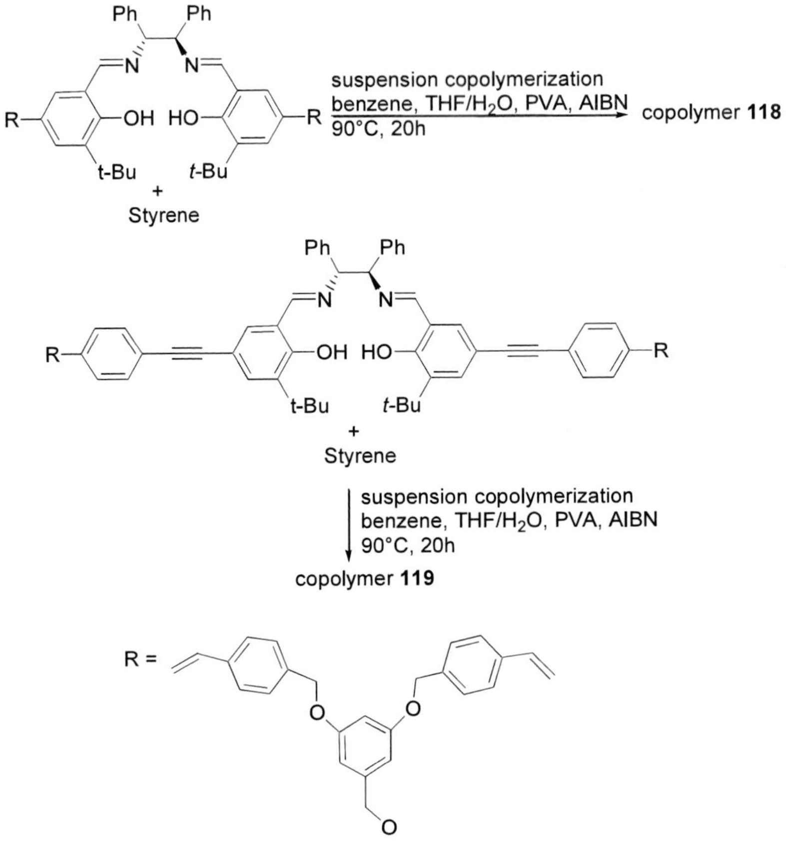

Scheme 60.

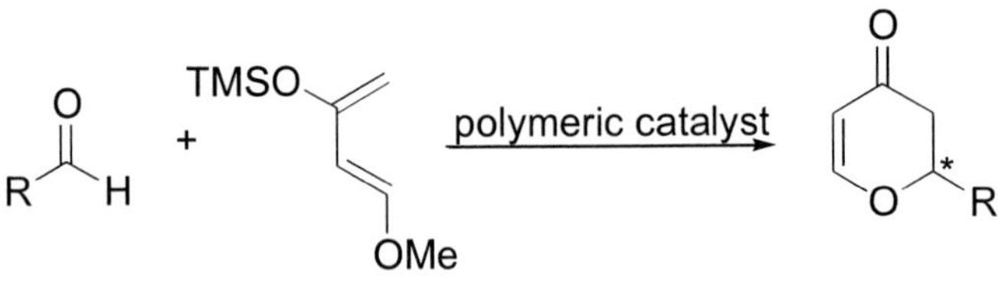

Scheme 61.

Compared to the homogeneous catalytic system analog, heterogeneous one is less selective. The difference in enantioselectivity between homogeneous catalyst and these electropolymerized ones were possibly due to the steric hindrance around the active site containing the metal and the electronic effects providing by the delocalization due to the conjugaison of the polymerized catalyst. Seven or fifteen recycles of the catalytic system have been performed with heptanal and cyclohexane carbaldehyde respectively without showing any change in enantioselectivity and activity.

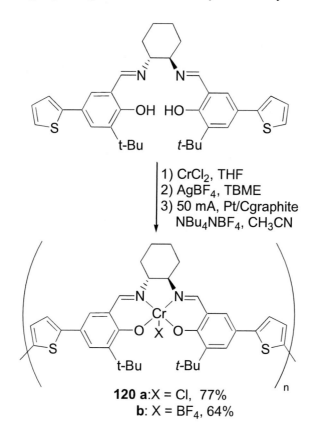

1) CrCl$_2$, THF
2) AgBF$_4$, TBME
3) 50 mA, Pt/Cgraphite
   NBu$_4$NBF$_4$, CH$_3$CN

120 a: X = Cl, 77%
b: X = BF$_4$, 64%

Scheme 62.

Chapter 5

ADDITION OF ORGANOMETALLIC REAGENTS TO ALDEHYDES, KETONES AND IMINES

The carbon-carbon bond forming reaction is one of the most useful chemical processes for the construction of complex natural or synthetic organic molecules. Addition of organometallic reagents to carbonyl or imine compounds is among the most fundamental reaction [111]. Its enantioselective version producing simultaneously a carbon-carbon bond and a chiral center is particularly important. The reaction of Grignard reagents with carbonyl compounds proved to be one of the best methods for forming carbon-carbon bond. Unfortunately, Grignard reagents are unsuitable for asymmetric alkylation because of their high reactivity in opposite to organozinc reagents such as diethylzinc, diphenylzinc or alkynylzinc which are excellent nucleophiles in the presence of chiral ligands (Scheme 63).

Numerous studies about the asymmetric addition of these reagents to carbonyl or imine derivatives have been performed and various chiral ligands have been synthesized [112]. Among the ligands used, chiral ligands bearing nitrogen chelating donor atom have been extensively studied due certainly to their easy preparation and availability. Furthermore, a lot of chiral ligands have proved to be highly efficient for the asymmetric alkylation of carbonyl or imine compounds in homogeneous phase.

Among the polymer-supported ligands, derived-aminoalcohol ones have been largely used in this reaction but few are derived either from *N*-sulfonamideaminoalcohols, or oxazoline or salen.

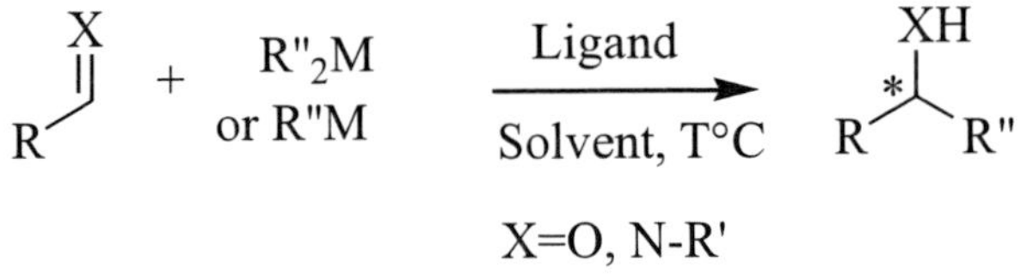

$$\underset{R}{\overset{X}{\parallel}} + \underset{or\ R''M}{R''_2M} \xrightarrow[\text{Solvent, T°C}]{\text{Ligand}} \underset{R \quad R''}{\overset{XH}{\underset{*}{\bigwedge}}}$$

$$X=O,\ N\text{-}R'$$

Scheme 63.

5.1. Supported Aminoalcohols as Ligands

The first work about polymer-supported β-aminoalcohol used in the asymmetric nucleophilic addition to aldehydes was reported by Fréchet [113]. The best result (95% ee) was obtained by using a polymeric catalyst derived from N,N-dialkylated (-)-3-exo-aminoisoborneol **124b** for the asymmetric addition of diethylzinc to o-methoxybenzaldehyde in toluene at 0°C (Scheme 64). This catalyst was prepared through reaction of aminoalcohols with 1-2% crosslinked chloromethylated polystyrene. The chiral polymeric catalyst of the reaction could be used several times in further asymmetric reactions.

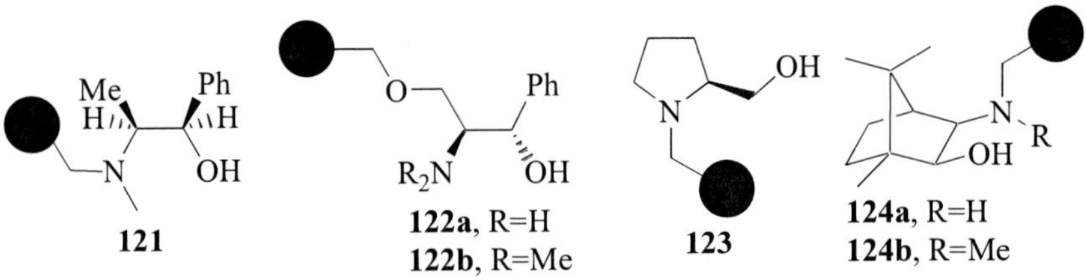

Scheme 64.

Soai has described the enantioselective addition of dialkylzinc reagent to aromatic and aliphatic aldehydes by using the supported-ligand **121** [114]. This latter was also prepared from (1R,2S)-(-)-ephedrine and chloromethylated polystyrene (1% divinylbenzene, 0.8 mmol Cl/g, 100-200 Mesh). In the case of benzaldehyde, the enantioselectivity was better than those reported with **124a** (89% vs 80%) and with the N-alkylephedrine-based homogeneous catalyst. The authors mentioned that the catalyst could be easily recovered and reused. They also studied the reaction with norephedrine homolog and reported enhanced enantioselectivity in the case of aliphatic aldehydes [115]. Polymer-supported ephedrine **121** was also employed in the first hetero-geneous enantioselective alkylation of N-diphenylphosphinoylimines and the best results (80% yield and 80 % ee) were obtained when the reaction was carried out in toluene at room temperature for 2 days [116].

Itsuno has reported interesting work about the effect of the difference of the structure of the polymeric part on the enantioselectivity of the reaction and the stability of the support [117]. They first prepared the chiral aminoalcohols **125** which could react with a crosslinked chloromethylated polystyrene or **126** which could be copolymerized with styrene and some styrene-derived crosslinking compound (Scheme 65).

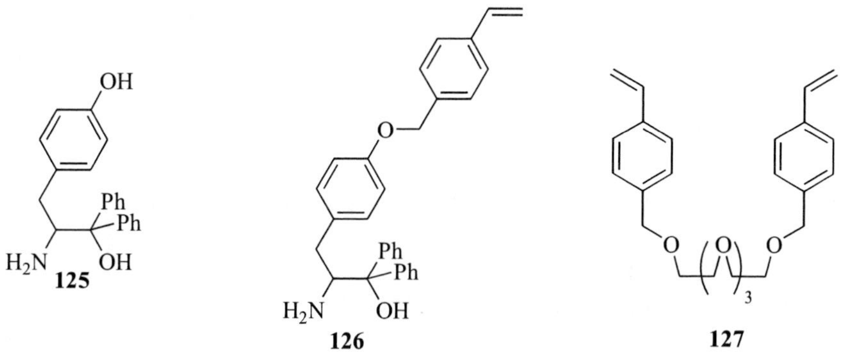

Scheme 65.

Five different chiral polymers containing the same aminoalcohol moiety were synthesized (Scheme 66). Reaction of aminoalcohol **125** with a chloromethylated polystyrene crosslinked with 2% of divinylbenzene led to chiral polymer **128**. When the highly cross-linked polystyrene containing 20% of DVB **129** was used, no reaction between **125** and this resin occurred because of the lack of the swellability of the resin in organic solvents. To avoid this problem, the crosslinking agent containing an oxyethylene chain **127** and the corresponding crosslinked chloromethylated polystyrene **131** were prepared. This latter swelled very well in organic solvents such as toluene and THF, even in the case of a highly crosslinked resin. Phenoxide anion of **125** could react with the chloromethyl group of polymer **131** to give the chiral supported ligand **132**.

Three other chiral polymers have been synthesized by copolymerization of the chiral monomer **126** with styrene in the presence or not of crosslinking agent (Scheme 66). Chiral polymer **133** was prepared by suspension copolymerization of aminoalcohol **125** with styrene and DVB in a 1/7/2 molar ratio. Polymer **133** swelled well in organic solvents such as toluene, benzene and THF and that unusual swellability could be due to the bulkyness and polarity of the chiral monomer **125**. The replacement of DVB by **127** as the crosslinking agent afforded chiral polymer **134** after copolymerization. This one also swelled well in organic solvents and presented interesting mechanical stability. The solution polymerization of **127** and styrene led to the soluble linear chiral polymer **135**.

The main problem of polymers containing a high degree of crosslinking is their stability with stirring over a long period of time. Indeed, their spherical beads are not enough stable and their breakdown during stirring lead to formation of a fine powder of the insoluble polymer. So the filtration at the end of the reaction is difficult and the reuse of the chiral polymer was not possible. The authors studied the effect of the magnetic stirring on the three highly cross-linked polymer beads and found that chiral 20% **127**-crosslinked polymers **132** and **134** were more stable than polymer **133** prepared by using 20% DVB. For example, when stirring at 350 rpm for 4 days in toluene, 50% of polymer **133** was transformed into a fine powder contrary to **132** and **134** which kept their original bead form.

The five chiral polymers **128** and **132-135** were then tested in asymmetric addition of diethylzinc to benzaldehyde. Good enantioselectivities were obtained with chiral polymers **133-135** in opposite to the two others polymers **128** and **132**. In the case of *p*-chlorobenzaldehyde, an excellent ee of 99% was obtained with polymer **134** in opposite to *o*-alkoxybenzaldehydes (21-54%) and longer reaction times were necessary to have satisfactory conversions with aliphatic aldehydes.

The recyclability of the chiral polymers was studied in a batch system using a simple loading of polymeric ligand **134** and both yield and ee were highly reproducible. Large scale of a chiral product could be obtained by using a continuous flow system. This technique have been tested with chiral polymer **133** and showed that a column containing 5 mmol of the polymer could produce about 90 mmol of (*S*)-1-(*p*-chlorophenyl)propanol with 94% ee. The main advantage of this technique is that it avoids the stirring which destroy the polymer during repeated reactions.

Soai and Watanabe have studied the effect of the introduction of a spacer between the ligand and the polymeric support [118]. In fact, when the polymer is directly connected to the nitrogen atom of the aminoalcohol, the polymer part could have an influence on the enantioselectivity since it is known that nitrogen atom is strongly involved in the formation of intermediate chiral zinc complex.

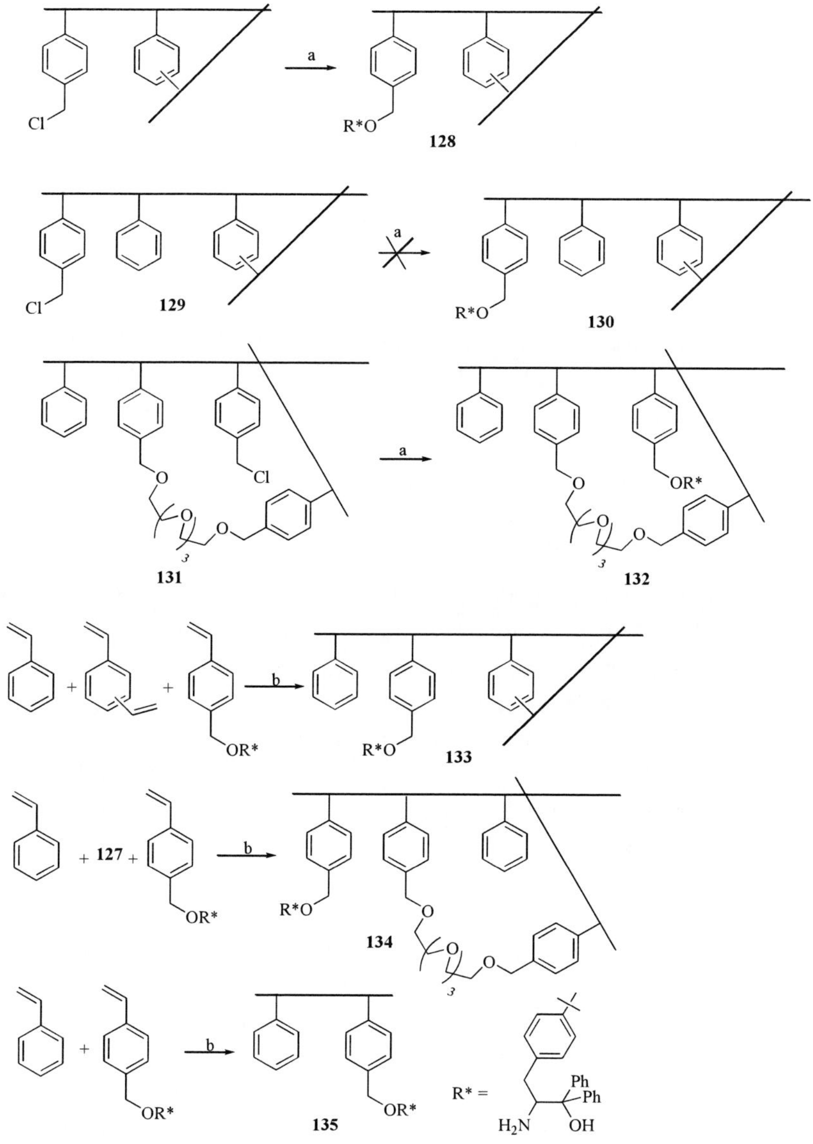

Scheme 66. (a) R*OH, NaH, DMF. (b) AIBN, THF.

The two chiral polymers **137** (content of aminoalcohol moiety is up to 0.085 mmol/g) and **138** (content of aminoalcohol moiety up to 0.25 mmol/g) were prepared from polymer **136** (1% DVB; Cl: 0.8 mmol/g; 100-200 mesh) and the corresponding chiral aminoalcohols (Scheme 67).

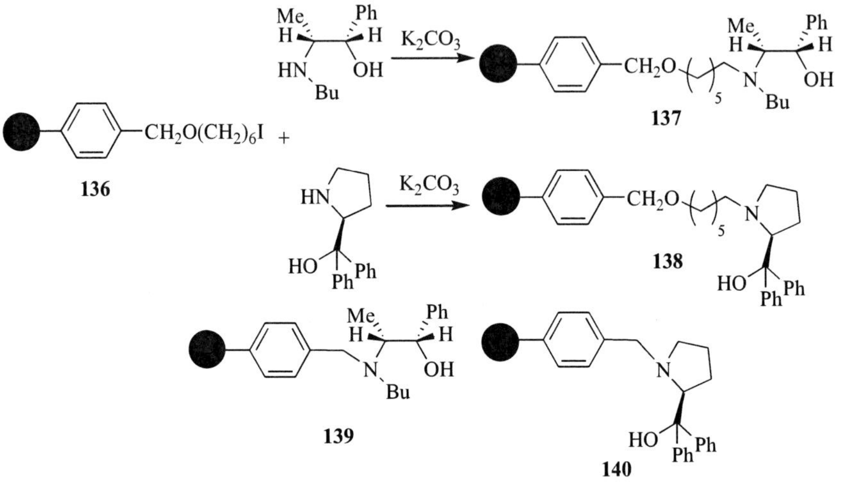

Scheme 67.

The enantioselective addition of diethyzinc to aldehyde at 0°C with chiral polymers **137-140** was examined. With aromatic and aliphatic aldehydes, higher enantioselectivities were obtained with chiral polymers **137** and **138** containing a six-methylene spacer compared to the two chiral polymers **139** and **140**. The same level of enantioselectivity was found with recycled chiral polymers. The authors assumed that the spacer has here two effects. The steric repulsion between the aminoalcohol part and the polystyrene is reduced and the spacer acts as a subtituent of the nitrogen atom which assists the stereochemical control of the reaction.

The four-methylene chain analog of chiral polymer **137** and chiral polymer **139** were also tested in the asymmetric alkylation of *N*-diphenylphosphonylbenzaldimines [119]. Contrary to aldehydes, both yield and enantioselectivity were lower with chiral polymer containing a spacer between the polymeric moiety and the catalytic part. The authors then examined the reaction conditions with **139**. The result of their screening was that the best yields and ee's were obtained when the reaction was carried out in toluene at room temperature in the presence of 1 equivalent of the polymeric ligand.

Ephedrine or norephedrine-derived copolymers **142a-d** have also been used in the same reaction [120]. These chiral polymers were prepared by copolymerization of monomer **141a-d** with styrene and DVB in a molar ratio **141a-d**/styrene/DVB of 1/7/2 (Scheme 68).

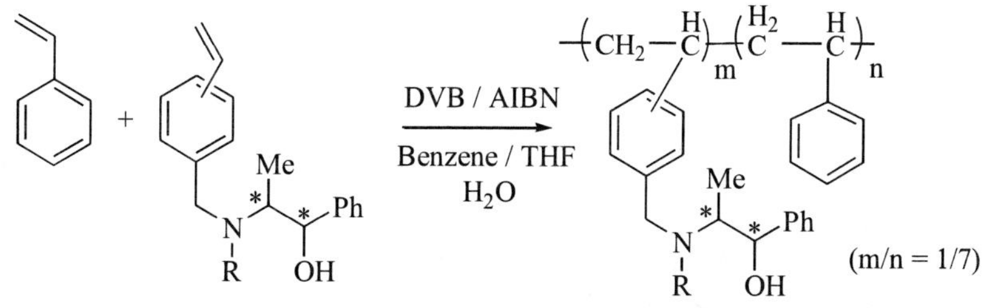

$$\left(CH_2-\overset{H}{\underset{}{C}}\right)_m\left(\overset{H_2}{\underset{}{C}}-\overset{H}{\underset{}{C}}\right)_n$$

DVB / AIBN

Benzene / THF

H_2O

(m/n = 1/7)

141a-d

(1S,2R)-142a: R=H   **(1S,2R)-142c**: R=Et
(1R,2S)-142b: R=Me   **(1S,2R)-142b**: R=Bu

Scheme 68.

The asymmetric addition of diethylzinc to N-diphenylphosphinyl-benzaldimine was carried out in toluene at room temperature in the presence of the chiral copolymerized ligands **142a-d**. Although the yields were low to moderate, the enantioselectivity was moderate to high (64%-88% ee) particularly with the chiral ligands **142b** and **142c**. The enantioselectivities with these two polymers were comparable with those observed with the corresponding monomer **141b** and **141c**. The recyclability of polymer **142b** was also tested and this one could be reused without loss of the enantio-selectivity.

Hosoya has reported the preparation of polymer-supported ligand which has only moderate enantioselectivity because the aim of the work was to show the effect of the type of immobilization [121]. The polymeric ligands were prepared by two different techniques, the Mix Method and the Add Method. The Mix Method was a classical polymerization: the ligand-containing monomer was dissolved in the mixture of monomers and porogenic solvent was used for the second-step swelling. After completion of this step, the polymerization step was carried out at 70°C with slow stirring for 24h to lead to the resulting polymers **144** and **145**. The Added Method consisted in adding the same amount of monomers **143** in the aqueous polymerization medium 4h after the initiation of the polymerization to give the polymers **146** and **147** (Scheme 69). The two monomers were quantitatively incorporated in the beads produced by the Mix Method and only 44% of monomers for the Add Method.

The catalytic activity of the different polymeric ligands was evaluated in the asymmetric ethylation of aldehyde in hexane. The polymers derived from the Mix Method gave better results in terms of activity and enantioselectivity than those from the Add Method. However, both yields and ee were slightly lower than those obtained with the monomeric ligands **137**.

When a mixture of toluene/hexane (1/1) was used as solvent, chemical yield dropped significantly.

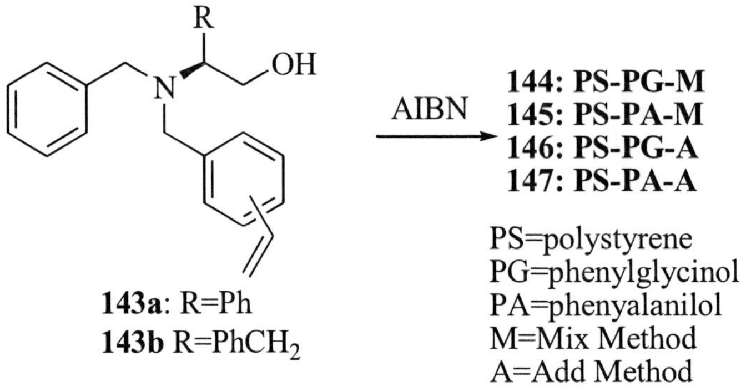
R
OH
N
AIBN
143a: R=Ph
143b R=PhCH2
144: PS-PG-M
145: PS-PA-M
146: PS-PG-A
147: PS-PA-A
PS=polystyrene
PG=phenylglycinol
PA=phenyalanilol
M=Mix Method
A=Add Method

Scheme 69.

The effect of the level of functionalization and crosslinkage of different Merrifield resins was also tested [122]. Chiral polymers **150** were prepared in two steps from enantiopure epoxyalcohol **148** (Scheme 70).

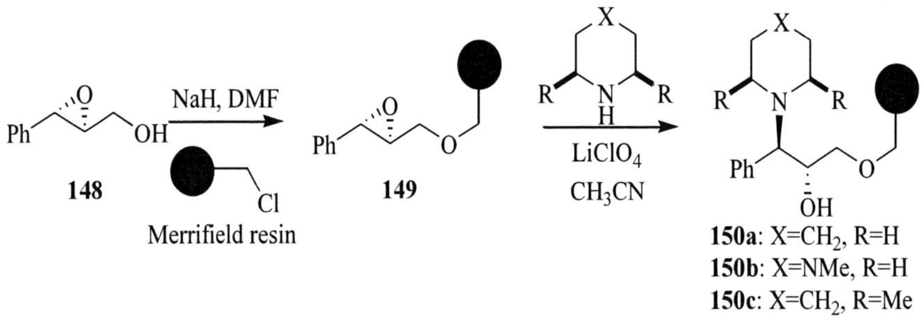
Ph
O
OH
148
NaH, DMF
Cl
Merrifield resin
149
X
R
N
H
R
LiClO4
CH3CN
OH
150a: X=CH2, R=H
150b: X=NMe, R=H
150c: X=CH2, R=Me

Scheme 70.

These chiral polymers **150** with different levels of crosslinking and ligand funtionalization were evaluated in the asymmetric addition reaction of diethylzinc to benzaldehyde and the results are summarized in Table 7. Although the conversions were higher than 95%, the enantioselectivities were low to moderate and significantly lower than those observed with the corresponding homogeneous ligand [123]. In all cases, the best results were obtained with both lower level of cross-linking (1%) and level of chlorine per gram of resin (1.2 mmol) of the chroromethylated polystyrene. Higher the level of aminoalcohol residue is, more important the formation of dimeric zinc alkoxide is. These dimeric species are suspected to be inactive in catalysis, requiring a preliminary dissociation before the catalysis step.

It was assumed that the low enantioselectivity was probably due to a lack of bulky susbstituents on the primary hydroxy group of the chiral aminodiol. The *cis* 2,6-dimethylpiperidine-derived aminoalcohol 151 was anchored to 2-chlorotrityl chloride resin (Barlos resin) leading to polymeric ligand 152 (Scheme 71).

Table 7. Enantioselectivity for the ethylation of benzaldehyde with chiral polymers 150

Ligand (%DVB ; f)	Amount (%)[b]	ee (%)
150a (1; 0.8)	5	36
150a (2; 1.6)	4	22
150a (2; 1.8)	3	20
150b (1; 0.9)	5	39
150b (2; 1.6)	4	20
150b (2; 1.7)	3	19
150c (1; 1.0)	3	69
150c (2; 1.6)	4	57
150c (2; 1.7)	3	57

[a] f=mmol of ligand/g resin (calculated by elemental analysis of nitrogen). [b] amount of catalyst (% molar with respect to benzaldehyde).

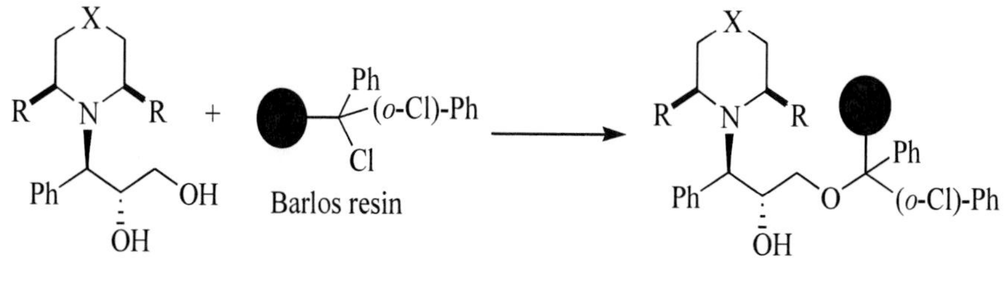

151   **152**

Scheme 71.

The use of an amount of 5 mol% of functionalized Barlos resin **152** in the asymmetric addition of Et$_2$Zn to benzaldehyde in toluene was highly efficient and (S)-1-phenylpropanol could be isolated with an excellent ee (entry 2, Table 8. When the reaction was carried out at 0°C, the enantioselectivity was slightly improved (entry 3, Table 8).

**Table 8. Asymmetric addition of diethylzinc
to benzaldehyde using 152**

Entry	f (mmol ligand/g of resin)	% molar ligand	T (°C)	Ee (%)
1	0.9	4	rt	79
2	1.1	5	rt	92
3	1.1	5	0	93
4	1.2	5	0	94

The asymmetric addition of Et$_2$Zn to different aromatic and aliphatic aldehydes using the functionnalized Barlos resin **146** was also studied and both conversions (> 82%) and ee's (>86 %) were high.

Hodge has also studied the reaction by using chiral polymers from ephedrine-derived monomer **141b** [124]. Homopolymerization of monomer **141b** in presense of AIBN in toluene and copolymerization with styrene or 4-methylstyrene in various proportions led respectively to chiral polymers **153**, **154** and **155** (Scheme 72).

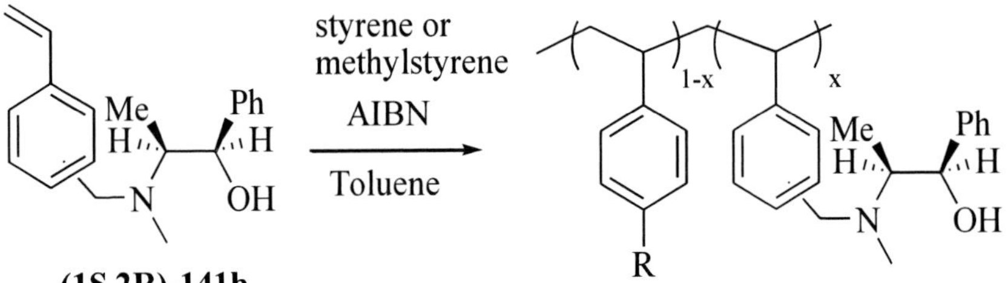

(1S,2R)-141b

153: x=1
154: R=H, x=0.01-0.66
155: R=CH$_3$, x=0.08-0.52
156: R=H, x=0.06-0.51; prepared
from crosslinked polymers beads

Scheme 72.

Chiral polymers **153-155** were used for the asymmetric addition reaction of diethylzinc to benzaldehyde. The reactions were carried out in toluene at room temperature for 24 hours. The results are summarized in Table 9. Although the major product was the 1-phenylpropanol, benzyl alcohol was detected as a minor product with each polymer **153** (9%), **154** (0-6%) and **155** (6-7%). As we can noticed in Table 9, the higher the loading of catalyst is, the lower the yield and the ee are. It can be explained for highly loading polymers by the presence of dimeric zinc alkoxide species which drop the catalytic activity of the catalyst. Higher yield and ee were obtained with Zn-complex derived from **154f** and **154g** which were soluble in toluene (entries 7-8, Table 9). In the case of polymers **155** (entries 9-13, Table 9), the reaction was heterogeneous and yield and ees increased as the loading decreased. The enantioselectivities obtained with polymers **154d-154g** (loading $\leq$ 1.32 mmol.g^{-1}) and **154d-155e** (loading $\leq$ 1.32 mmol.g^{-1}) were similar or better than those obtained with the model ligand, (1R,2S)-N-benzylephedrine.

The authors then turned their attention to the preparation of crosslinked polymer beads which are more insoluble and are more convenient to use than the linear polymers **153-155**. Different chiral crosslinked polystyrene beads **156** were prepared by reaction of various crosslinked polystyrene beads with (1R,2S)-N-benzylephedrine in the presence of potassium carbonate. The catalytic properties of these various crosslinked polymer-supported ephedrine have been tested in the asymmetric addition reaction of diethylzinc to benzaldehyde and the results are summarized in Table 10. The reactions were carried out either in toluene or hexane at 0°C or 20-23°C. As it can be noticed in Table 10, the enantioselectivities were higher in toluene. Compared to linear polymers **153-155**, lightly crosslinked chiral polymers led to the best results in terms of yield and ee. The use of Amberlite XAD-4 and Polyhipe™ as support proved to be not efficient. It was assumed that probably for both supports the catalytic sites are more in the inner surfaces of the pores and that result in a high local concentration of aminoalcohol residue which lead to a significant site-site interaction. Chiral polymer **156a** was also used in the asymmetric ethylation of 2-methoxybenzaldehyde and 4-chlorobenzaldehyde under similar conditions and ees reached 90-93% when the reaction was performed at 0°C.

Three other polymers **158** were prepared by homopolymerization and copolymerization of aminoalcohols **157** and by grafting of the corresponding aminoalcohol onto 1% crosslinked polystyrene beads containing 0.70 mmol.g^{-1} of chloromethyl group leading to a polymer containing 0.64 mmol.g^{-1} of ligand residue (Scheme 73) [124].

Contrary to ephedrine-derived polymers **153-156**, the use of a high loading of the catalytic residue ($\geq$ 2.48 mmol.g^{-1}) for chiral polymers **158a** and **158b** gave excellent enantioselectivies for the ethylation of benzaldehyde and similar to those obtained with the corresponding non-polymeric chiral aminoalcohol.

**Table 9. Effect of the ligand loading of polymers
153-155 on the catalytic activity**

Entry	Polymer	Fraction of monomer **135b** (x)	Loading (mmol.g^{-1})	Yield (%)	ee (%)
1	**153**	1	3.52	47	44
2	**154a**	0.66	2.98	56	56
3	**154b**	0.52	2.65	66	67
4	**154c**	0.28	1.81	80	79
5	**154d**	0.18	1.32	95	81
6	**154e**	0.05	0.44	98	82
7	**154f**	0.03	0.27	100	83
8	**154g**	0.01	0.09	100	83
9	**155a**	0.52	2.56	65	69
10	**155b**	0.46	2.38	67	74
11	**155c**	0.36	2.03	74	76
12	**155d**	0.17	1.16	92	81
13	**155e**	0.08	0.61	92	86

**Table 10. Asymmetric diethylzinc addition to
benzaldehyde using crosslinked polymer beads 156**

Entry	Polymer	Starting polymer	Catalyst loading	Degree of substitution	Yield (%) (yield)^a	ee (%) (ee)^a
1[b]	**156a**	1% crosslinked gel	0.93	0.12	81 (81)	81 (64)
2[c]	**156a**	1% crosslinked gel	0.93	0.12	85 (-)	78 (-)
3[b]	**156b**	1% crosslinked gel	1.13	0.15	77 (83)	77 (62)
4[c]	**156b**	1% crosslinked gel	1.13	0.15	85 (81)	74 (65)
5[b]	**156c**	1% crosslinked gel	2.30	0.40	80 (82)	74 (62)
6[c]	**156d**	1% crosslinked gel	2.30	0.40	86 (97)	73 (65)
7[b]	**156d**	2% crosslinked gel	1.06	0.14	70 (-)	72 (-)
8[b]	**156e**	2% crosslinked gel	2.62	0.51	71 (76)	69 (62)
9[b]	**156f**	Amberlite XAD-4	0.90	0.11	59 (62)	28 (39)
10[c]	**156g**	Amberlite XAD-4	0.50	0.06	75 (70	28 (39)
11[b]	**156h**	Polyhipe™	1.51	0.21	59 (57)	42 (23)

[a] Under brackets, reaction was carried out in hexane.
[b] Reaction was carried out at 0°C for 70h.
[c] Reaction was carried out at 20-23°C for 20h.

This surprising excellent result with a high loading of aminoalcohol unit polymer is probably due to their high solubility in toluene. The enantio-selectivity obtained with the crosslinked polymer beads **158c** with a ligand loading of 0.64 mmol.g^{-1} was also high (97% ee).

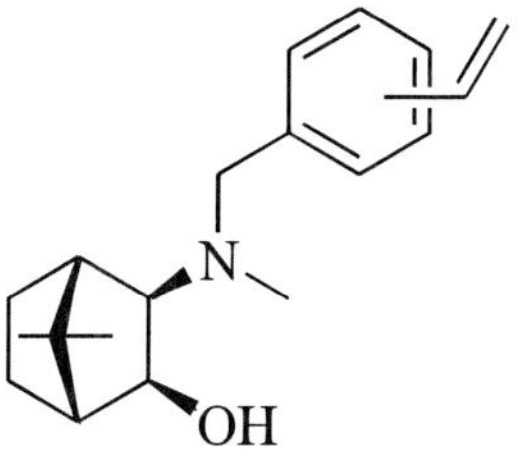

157

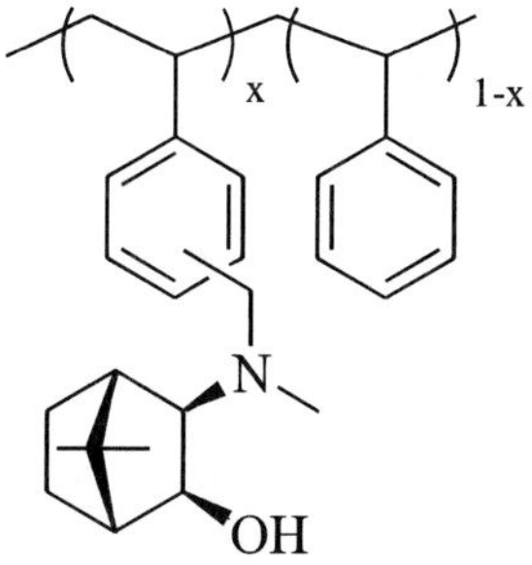

158a: x=1
158b: x=0.66
158c: x=0.11; prepared from
crosslinked polymers beads

Scheme 73.

The authors were then interested to use the two enantiomers of ephedrine-
and camphor-derived polymeric ligands **153-156** and **158** in a flow system
[125]. When polymeric ligand **156a** and its enantiomer analog (1%
crosslinking, 1.11 mmol/g of aminoalcohol residue) were employed with a
flow rate of 2 mmol.h^{-1} of benzaldehyde and a diethylzinc/benzaldehyde ratio
of 2.5/1, the ee were respectively 81% and 72%. Lightly crosslinked beads **156**
(0.2% crosslinking, 1.78 mmol/g loading) produced the (*R*)-alcohol in a yield
of 98% and in 98% ee. With a flow rate of 2.5 mmol.h^{-1} of benzaldehyde, the
ee dropped slightly (91% *vs* 98%).

Supported-ligand **158c** (with a ligand loading of 0.64 mmol/g) was also
tested in a flow system because this polymer-supported catalyst gave better
results than ephedrine one **156**. A loss of catalyst performance was noticed
here; indeed, the chemical yields and enantioselectivity dropped significantly
after aproximatively twelve runs (run 1: 95% yield and 97% ee; run 12: 59%
yield and 87 % ee). It was suggested that the deterioration in performance
could be due to the oxidation of the hydroxyl group of the aminoalcohol unit.

(*S*)-aziridinyldiphenylmethanol **159** was immobilized on polymer-bound
triphenylchloromethane (1.1 mequiv/g) to lead to the new chiral polymer **160**
(Scheme 74) [126]. This latter was employed in the enantioselective addition
of diethylzinc to aromatic and aliphatic aldehydes. For example, (*S*)-1-
phenylpropanol was isolated in a yield of 92% and an ee of 96% when the
reaction was performed in a 1/1 mixture of toluene and dichloromethane. The
presence of dichloromethane proved to be essential for swelling of the
polymer and making the catalytic sites accessible to substrates. With other

aromatic aldehydes and cyclohexane carboxaldehyde, the same range of chemical yield and ee was obtained. But with acyclic aliphatic aldehydes the chemical yield and ee dropped significantly. The recyclability of the chiral polymer **160** has been investigated and it was noticed a slightly decrease in enantioselectivity (1% per run).

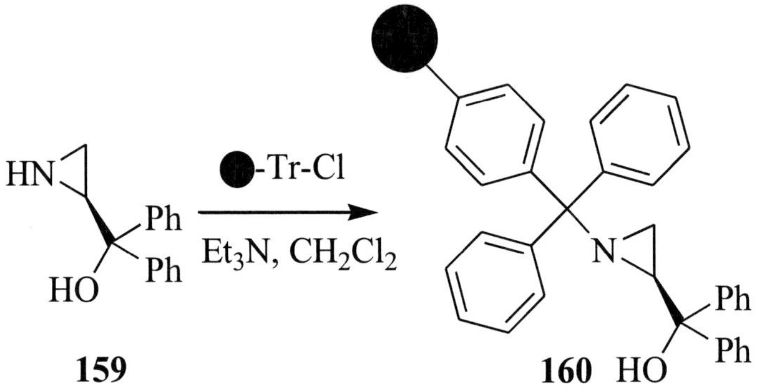

Scheme 74.

Styrenic L-prolinol-derived ligand *N*-Boc-**60** was immobilized on chemical inert, mechanically stable polyethylene fibers by electron beam induced preirradiation grafting using styrene as a comonomer to lead to polymer fiber **161** with a loading of 0.2 mmol/g [127]. Carbamate function was then removed under classical conditions to produce polymer **162** (Scheme 75). Chiral polymer **162** was used as ligand in the diethylzinc addition to benzaldehyde and yield and ee were very low (36% and 24% respectively). A higher yield of 55% could be obtained in a titanium-mediated addition of diethylzinc to benzaldehyde by using the protected chiral polymer **161** with the same enantioselectivity.

In order to improve the catalytic properties of this supported-α,α-diphenylprolinol polymer fibers, Degni has tested the immobilized ligand **67** with a loading of 0.4 mmol/g (Scheme 30) [128] in the diethylzinc addition to benzaldehyde. The chemical yields were largely superior to those obtained with polymer **162** (95-98% *vs* 30-36%) and the enantioselectivities were also higher and reached 40% which was comparable to the corresponding non-polymeric ligand.

Chiral polymer **67** was also tested in another enantioselective reaction, the Carreira reaction. The phenylacetylene addition to benzaldehyde in the presence of zinc triflate using **67** as polymeric ligand was carried out. The corresponding propargyl alcohol was isolated in a moderate yield but with an excellent enantioselectivity of 91%. This supported ligand was also recycled for this reaction whithout loss of catalytic activity.

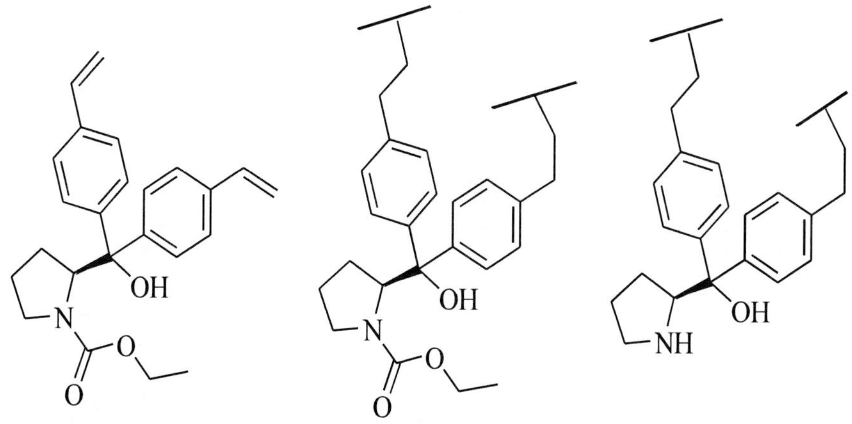

N-Boc 60   **161**   **162**

Scheme 75.

A new strategy for optically active chiral polymers was envisaged and based on a Suzuki coupling polymerization of (1R,2S)-ephedrine unit-bearing monomer with rigid linkers [129]. The Suzuki coupling of ephedrine-derived dibromide **163** (which was prepared in four steps from (1R,2S)-ephedrine) with the diboronic acid **164** in the presence of Pd(PPh$_3$)$_4$ afforded chiral polymers **165** (Scheme 76). Synthesis of chiral polymer **167** containing a longer linker between two chiral units has been performed by using diboronic acid **166**. Both polymers were soluble in common solvents such as CH$_2$Cl$_2$, THF, CHCl$_3$...

The chiral polymeric ligands **165** and **167** were then employed in the diethylzinc addition to benzaldehyde and anisaldehyde. The enantio-selectivities obtained with both polymers were between 74 and 76% for the two aldehydes The polymers could be recovered by precipitation from methanol and the recycled polymeric ligand showed the same catalytic properties.

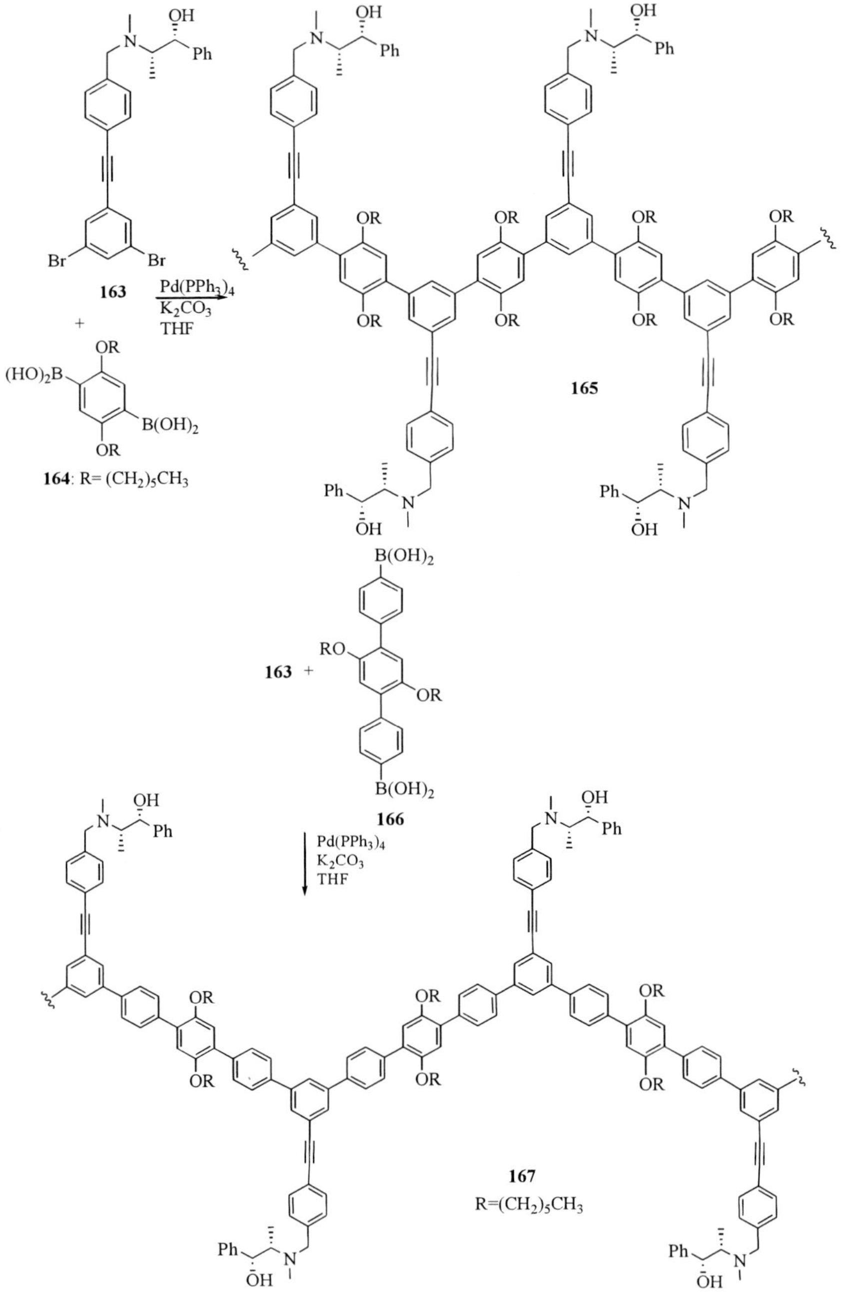
163
Pd(PPh3)4
K2CO3
THF
164: R= (CH2)5CH3
165
163 +
166
Pd(PPh3)4
K2CO3
THF
167
R=(CH2)5CH3

Scheme 76.

Chiral bicyclic aminoalcohol-containing polymers **170** and **173** were produced from the industrial waste material **168a** generated in the industrial process of synthesizing the ACE inhibitor Ramipril at Aventis. Supported ligand **170** was obtained by grafting of compound **168b** (obtained in two steps from **168a**) onto a Merrifield resin (1.1mmol Cl/g) and after reaction of the resulting polymer with phenyl Grignard (Scheme 77) [130]. Excellent yields and ee (83 and 89 % respectively) were observed for the diethylzinc addition to benzaldehyde for the heterogeneous ligand **170** and were similar to those obtained for the homogeneous reaction.

In a second time, the authors turned their attention on the preparation of flow systems. They considered that in some cases the chiral supported ligands isolated by polymerization led to higher enantioselectivity. The monolithic column was selected as polymerization technique because the resulting material is more suitable for the preparation of flow systems because of its better porosity properties. Monoliths **173** with the desired morphology and properties were obtained by mixing 40 wt% of monomers (10 mol% of **172**/90 mol% of DVB; no styrene) and 60 wt% of toluene/1-dodecanol as the porogenic mixture (10 wt% of toluene). A flow system was performed and after 24 h, (*R*)-1-phenylethanol was isolated with an excellent ee of 99%. This enantioselectivity is better than those obtained with the corresponding homogeneous ligand and with the grafted ligand **170** and is one of the best ee observed with a polymeric ligand. The authors assumed that this efficiency could be due to the formation of more appropiated chiral cavities because of the polymerization process or to the high level of crosslinking agent used which isolates the aminoalcohol residue. Chiral polymer **173** could be reused for four successive runs without loss of the enantioselectivity.

Reduction of the ester function of polymer **169** or treament with a dimagnesium bromide reagent led to the two immobilized aminoalcohols **174** and **175** (Scheme 77). Chiral polymer **174** proved to be not efficient for the diethylzinc addition to benzaldehyde (11% yield and no ee) and with chiral polymeric ligand **175**, for the same reaction, the yield was excellent but the enantioselectivity was low (40%) [131].

The same group has developped a small library of supported chiral aminoalcohols from different amino acid [132]. The hydrochloride salt of different amino acid methyl esters **176** reacted with a chloromethylated polystyrene-divinylbenzene (1 mmol Cl/g, 1% DVB) under basic conditions to afford the polymeric amino acid ester **177**.

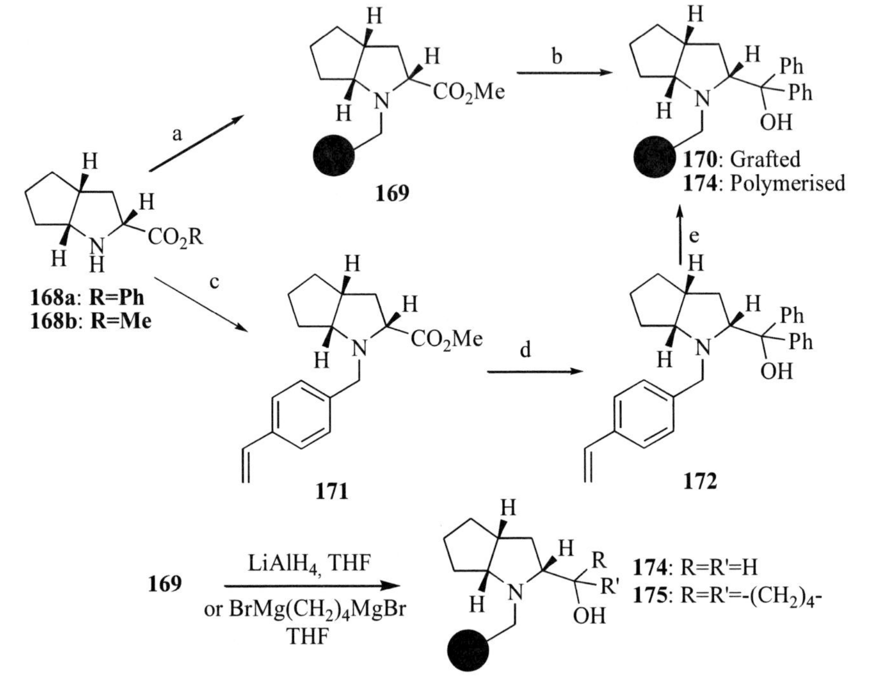

Scheme 77. Reagents and conditions: (a) Merrifield resin (1.1mequiv/g, 1%cross-linking), NaHCO₃, THF. (b) PhMgX, THF. (c) 4-vinylbenzyl chloride, NaHCO₃. (d) PhMgCl, THF. (e) A49/DVB/Toluene/1-dodecanol, AIBN.

Reduction of the ester function or addition of Grigrard reagents led to supported-amino alcohols **178-181** (Scheme 78). The nitrogen atom of these polymers was then methylated to afford the *N*-methylaled polymeric aminoalcohols **182-185**.

Chiral polymers **178-181** were first tested for the diethylzinc addition to benzaldehyde using 10 mol% of the supported catalyst. In all cases, the chemical yield and the selectivity (based on the 1-phenylethanol/benzyl alcohol ratio) were moderate and the ee were around 10 % except for the (*S*)-proline-derived polymer **180e** (45% ee). The low catalytic activity could be due to the presence of the NH group in polymers **178a-d – 181a-d**.

In order to improve the catalytic activity of these suppported ligands, the nitrogen atom of these polymers was then methylated to afford the *N*-methylaled polymeric aminoalcohols **182-185**. The catalytic activity of these heterogeneous ligands was evaluated and the results are summarized in Table 11. The analysis of the results showed that an increase in selectivity is accompanied by an increase in the enantioselectivity. Polymers **182** were poor catalysts and that is probably due to the non-presence of substituent in α position. The most surprising results are those obtained with polymers **183** bearing two phenyl groups in α position since this type of aminoalcohols give generally high enantioselectivity. The best results were obtained with immobilized aminoalcohols **184a** and **185b** which are derived from respectively valine and leucine. This last one gave a higher selectivity (91 *vs* 86 % for **184a**) but a lower enantioselectivity (74 *vs* 80 % for **184a**) (entries 9 and 14, Table 11).

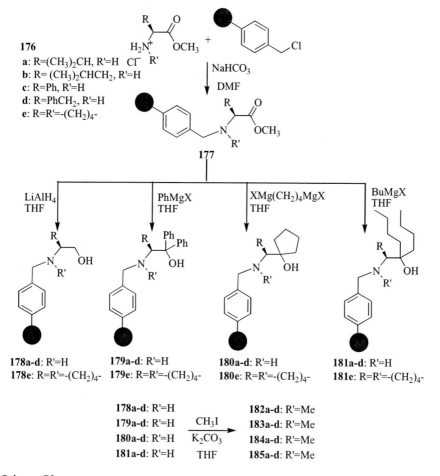

Scheme 78.

Supported chiral bispidine-derived aminoalcohols were screened in the addition of diethylzinc to benzaldehyde [133]. These immobilized ligands were obtained in three steps from compound **186** isolated from *N*-Boc-piperidin-4-one (Scheme 79). The hydroxyl function of **186** was first linked onto chloromethylated polystyrene to afford polymer **187**. The carbamate was then removed under classical conditions and reaction of the resulting polymer with chiral epoxides led to chiral polymeric bispidine-derived amino alcohols **188**.

**Table 11. Enantioselective diethylzinc addition
to benzaldehyde using 182-185**

Entry	Polymer	Yield (%)	Selectivity (%)	ee (%)
1	**182a**	89	55	10
2	**182b**	94	65	15
3	**182c**	93	56	7
4	**182d**	82	48	2
5	**183a**	89	63	15
6	**183b**	90	70	24
7	**183c**	83	58	11
8	**183d**	97	70	24
9	**184a**	99	86	80
10	**184b**	81	53	22
11	**184c**	90	68	23
12	**184d**	92	70	20
13	**185a**	92	73	34
14	**185b**	98	91	74
15	**185c**	94	86	17
16	**185d**	95	85	45

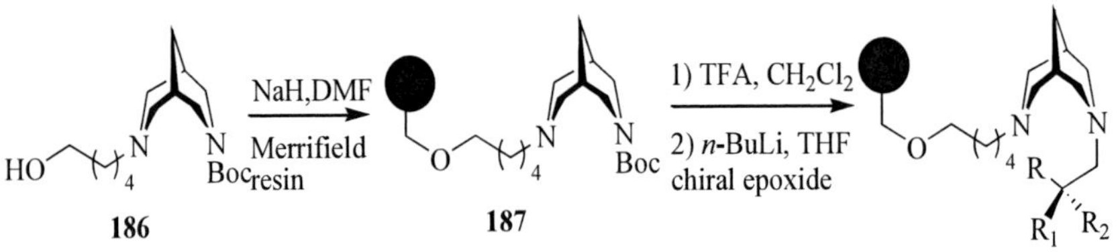

188a: R=Ph, R_1=OH, R_2=H
188b: R=Et, R_1=H, R_2=OH

Scheme 79.

Catalytic activity of both chiral polymers **188** was then evaluated in the addition of diethylzinc to benzaldehyde. Although yield reached 95%, the enantioselectivity was poor (25% ee for **188a** and 5% ee for **188b**) contrary to the corresponding non-polymeric ligand homolog of **188a** (96% ee).

Pericas has reported the immobilization of (2*R*)-2-piperidino-1,1,2-triphenylethanol analog **190** onto several resins with different levels of crosslinking and functionalization [134]. These different chiral supported aminoalcohols **191** and **192** were prepared by anchoring the aminoalcohol **190** on different Merrifield resin (0.63-1.49 mmol Cl/g; 1 or 2 % DVB) and on two Barlos resin (1.24-1.6 mmol Cl /g) (Scheme 80) *via* the benzylic hydroxyl group. These functionalized resins were then tested in the enantioselective diethylzinc addition to benzaldehyde and the results are summarized in Table 12. The five polymer-supported aminoalcohols exhibited high enantio-selectivities (>92 %) and except **192b**, they presented a similar profile of catalytic activity. After six consecutive runs, the recovered **191c** showed no loss of performance. This supported-ligand **191c** was also used with other aldehydes including aromatic and aliphatic aldehydes and both conversion and enantioselectivities were excellent (>95%).

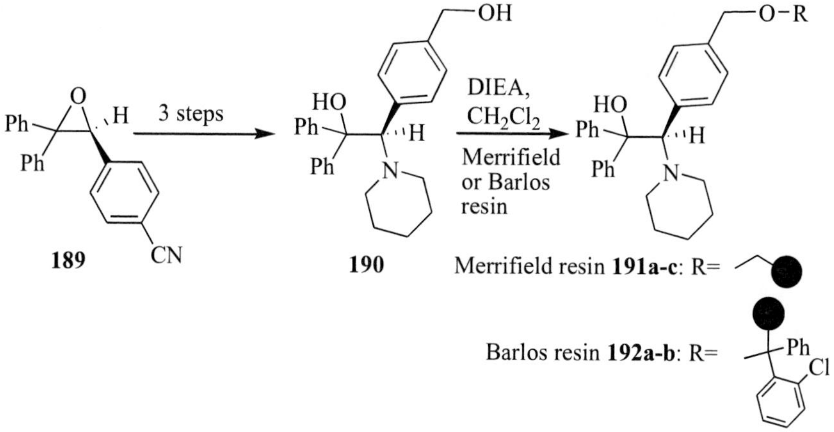

Scheme 80.

The asymmetric phenylation of different aldehydes was performed with chiral supported ligand **191** and the corresponding diarylmethanols were isolated in good to excellent yields (74-99%) and high enantioselectivities (85-91%) were observed but were slightly lower to those recorded with the homogeneous ligand [135].

**Table 12. Effect of functionnalized resins
191 and 192 on the enantioselectivity**

Entry	Starting resin	Functionalized resin (f^b)	Ee
1	Merrifield (1%DVB; f_0^a = 0.63)	191a (0.375)	95
2	Merrifield (1%DVB; f_0 = 1.49)	191b (0.850)	93
3	Merrifield (1%DVB; f_0 = 0.63)	191c (0.389)	95
4	Barlos (f_0 = 1.6)	192a (0.925)	93
5	Barlos (f_0 = 1.24)	192b (0.689)	92

[a] f_0=mmol Cl/g. B f=mmol/g (calculated by elemental analysis of nitrogen with the following formula: f=0.714% N).

Although the enantioselectivity obtained previously were among the best for supported ligands, the preparation of the aminoalcohol unit just before anchoring required up to five steps from commercial precursors. The supported-ligand **195** was prepared in only two steps from (S)-triphenyl-ethylene oxide **193** (Scheme 81) [136]. The aminoalcohol **194** was anchored to a Merrifield resin (2% DVB, f_0 = 0.84) with an anchoring yield of 97 %.

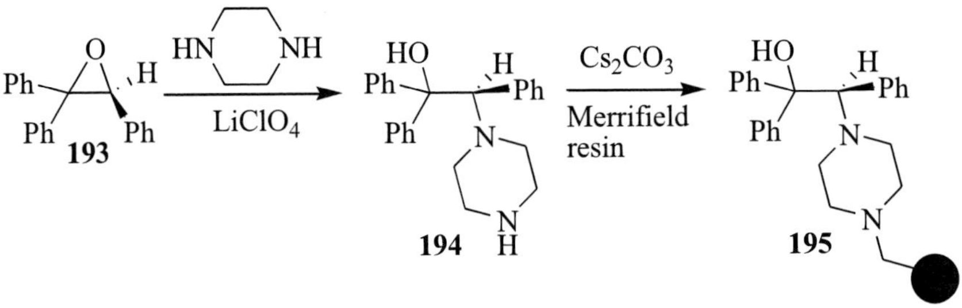

Scheme 81.

The catalytic activity of the polymer-supported aminoalcohol **195** was then evaluated for the diethylzinc addition of aldehydes. After optimization of the reaction (temperature, amount of catalyst), two types of reaction conditions (2 mol% of **195**, 0°C, 8h and 4 mol% of **195**, 0°C, 6h) were tested with 15 different aldehydes. With the first reaction condition type, the conversion and the enantioselectivity were already high ($\geq$ 95%) for most aromatic adehydes, the second one permit a complete conversion for all aromatic and aliphatic aldehydes and a better one for the α-substituted aldehydes.

In order to evaluate the effect of the polymer part on the catalytic activity, two parallel kinetic measurements of the diethylzinc addition to benzaldehyde using either **195** or the *N*-benzyl non-polymeric ligand were performed. The profiles for both ligands are pratically similar and at short reaction times, the polymer-supported ligand **195** showed a slightly better activity than its homogeneous analog (83% conversion *vs* 80% after 30 min).

This supported ligand proved to be highly recyclable in five consecutive runs without any loss of catalytic activity (> 99% conversion) and enantioselectivity (average of 95.1 %).

The enantioselective phenyl transfer to different aldehydes was also carried out using the chiral polymer **195** with 5 mol % of catalyst amount and the enantioselectivities were also high ($\geq$ 82 %) for aromatic aldehydes and moderate for aliphatic ones (68-79 %) [135].

Polymeric regioisomer of **191** and **195** was also synthesized and evaluated in the ethylation of aldehydes since the homogeneous analog was a highly active ligand [135]. Aminoalcohol **197** which was synthesized in six steps from epoxide **196** was anchored to Barlos resin with f_o = 1.6 to lead to polymeric ligand **198** (f = 0.90) (Scheme 82).

The ethylation of aromatic and aliphatic aldehydes in toluene at 0°C in the presence of an amount of 8 mol% of the supported-ligand **198** was performed. Except for α-susbituted aldehydes (around 50% conversion), both conversion and ee were high. The phenylation of aromatic aldehydes was also tested with polymeric ligand **198**. Although conversions were excellent, the enantioselectivities were moderate (38-48% ee) and lower than those obtained with functionalized resin **191** and **195**.

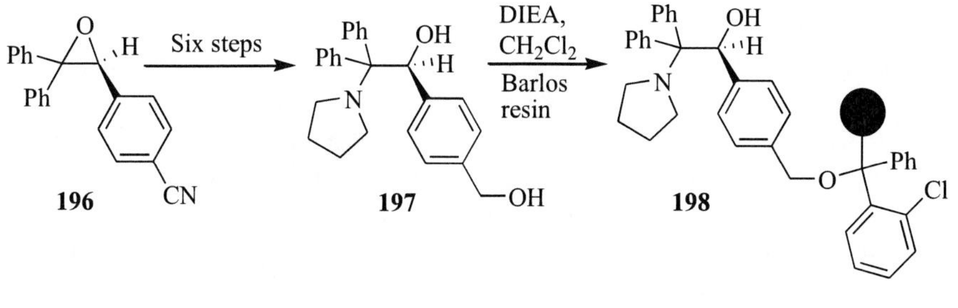

Scheme 82.

In order to avoid the limitation of the use of the anchoring step as a source of diversity to prepare libraries of supported ligands containing the same type of aminoalcohol as **195,** click chemistry was employed. This more convenient and efficient technique is a copper(I)-catalyzed cycloaddition between an alkyne and an azide [137]. Aminoalcohol-derived alkynes **199** were prepared by reaction of **194** with different bromine reagent. The reaction between these alkynes and a variety of azido resins with different degree of functionalization (f = 0.74 – 2.25 mmol N_3/g) was performed to afford the supported-ligands **200** (Scheme 83).

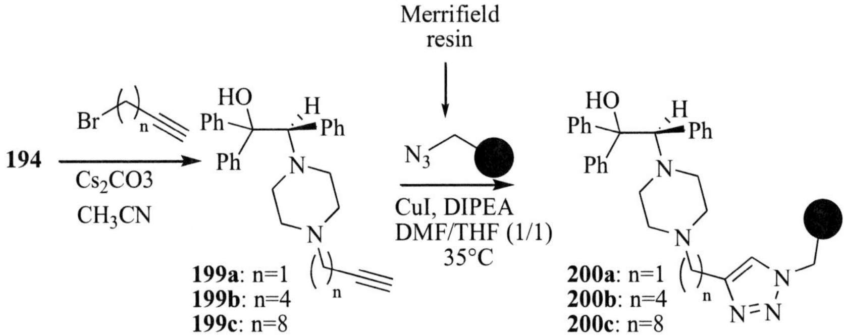

Scheme 83.

The new functionalized resin **200a** was first tested in the phenylation of *p*-tolualdehyde and *p*-anisaldehyde in toluene at 10°C using 5 mol% of the catalyst and the dependance of enantioselectivity on the level of functionalisation was studied (entries 1-6, Table 13). The first analysis of the results was that the enantioselectivity depended strongly on the level of functionalisation of the resin. For the phenylation of *p*-tolualdehyde (entries 1-4), the higher the degree of functionalization, the lower the enantioselectivity. When a high ligand-loaded polymer was employed, the diarylcarbinols were isolated in their racemic form (entry 4). In this case, the triazole ring could interfere with the catalytic site involving a drop of the enantioselectivity. In order to avoid that, catalytic activity of resins **200b** and **200c** bearing a spacer were evaluated in the phenylation of *p*-tolualdehyde and *p*-anisaldehyde (entries 7-14, Table 13). It can be noticed that the two polymeric ligands **200a** and **200b** gave the best results in terms of conversion and ee. Resin **200b** with a level of functionalization of 0.72 mmol/g exhibited the best catalytic activity (entries 9 and 13). This last one was chosen for phenylation of other aromatic and aliphatic aldehydes and the same range of conversion and enantio-selectivity was obtained. The recyclability of this supported-ligand was studied on *p*-anisaldehyde and no decrease of the conversion was observed after three consecutive runs while a slightly drop in the enantioselectivity was recorded.

The enantioselectivities were around 20% lower than those obtained with similar supported-ligand **195** with no triazole ring and for that reason the idea that the triazole ring could interact with zinc reagent is reinforced.

Table 13. Effect of the level of functionalisation
of polymer 200 on activity and selectivity

Entry	Polymeric ligand	f (mmol L/g)	Conversion (%)	ee (%)
1	**200a**	0.59	95	74
2	**200a**	0.74	>97	68
3[a]	**200a**	0.71	89	63
4	**200a**	1.16	>99	0
5	**200a**	0.59	64	77
6[b]	**200a**	0.71	92	71
7	**200b**	0.66	100	66
8[c]	**200b**	0.66	68	66
9	**200b**	0.72	95	76
10[b]	**200b**	0.72	99	78
11	**200c**	0.67	79	75
12	**200b**	0.66	68	77
13	**200b**	0.72	97	82
14	**200c**	0.67	54	76

[a] 3.6 % of catalyst.
[b] 10% of catalyst.
[c] Higher crosslinked polymer (2% DVB).

Until now, supported diphenylprolinol were build either by copolymerization with a vinylbenzyl group on nitrogen atom or a vinyl group on phenyl or by anchoring the ligand onto a Merrified resin *via* its nitrogen atom. Chen has reported a strategy to support this type of ligands using 4-hydroxy-diphenylprolinol **201** [138]. Basic treatment of **201** with polymers **202** prepared in two steps from Merrifield resin (1% DVB, 100-200 mesh, 1 mmol Cl /g) led to supported-aminoalcohols **203a** and **203b** with a loading of aminoalcohol residue of 0.66 and 0.42 mmol/g respectively (Scheme 84). .

These two immobilized ligands were tested in the diethylzinc addition to benzaldehyde and the best results were obtained when the reaction was performed in hexane at rt using an amount of 10 mol% of catalyst. Polymeric ligand **203a** showed a slightly better activity than **203b** (91 % yield and 63% ee *vs* 85% yield and 57% ee). The optimized reaction conditions in the presence of **203a** as ligand were used with different aldehydes. Except for salicylic aldehyde, *p*-nitrobenzaldehyde and cyclohexanecarboxaldehyde, the same range of yield and ee were observed. Five consecutive diethylzinc addition to benzaldehyde reactions using the same batch of polymer **203a** were carried out without loss of the catalytic activity for the three first ones while a slight decrease of the catalytic was detected for the fourth and fifth run.

The replacement of the ether bond between the 4-hydroxyprolinol and the spacer by an ester link has been performed [139]. Wang has prepared the polystyrene-supported diphenyl prolinol **205** and the soluble polymer-supported analog **206** in two steps from aminoalcohol **201** (Scheme 84).

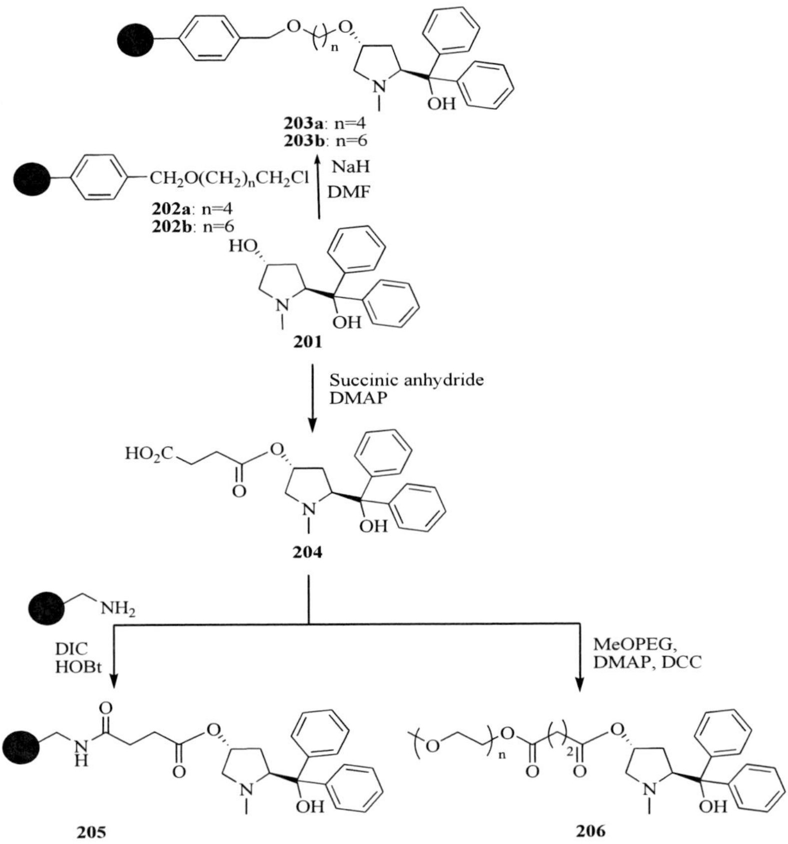

Scheme 84.

The loading of aminoalcohol was evaluated to be at 0.76 mmol/g for supported ligand **205** and 0.42 mmol/g for the soluble one. With polystyrene-supported ligand **205** for the enantioselective diethylzinc addition to aldehyde, the best result (92 % yield, 68% ee) was obtained when the reaction was carried out in toluene at 25°C using 15 mol% of the catalyst.

Both yield and ee were similar to those obtained with the ether-linked polymeric ligand 203 but the reaction time was shorter. The optimized conditions were also used with other aromatic aldehydes and the enantioselectivities were moderate to good (34-71%).

Soluble polymer-supported ligand 206 was also tested and for all the aldehydes tested both yield and enantioselectivity were higher than those using the polystyrene-supported 205. This latter could be reused at least 3 times without significant loss of performance contrary to its soluble analog.

The first immobilization of a pyridine-based tridentate ligand has also been investigated since the tridentate can stabilize the zinc complex and reduce the effect of the polymer blackbone [140]. The aminoalcohol 207 was prepared in six steps from 2,5-dibromopyridine and was anchored onto a 2% DVB crosslinked Merrifield resin to afford the supported-tridentate ligand 208 (Scheme 85). This immobilized ligand was then employed in the diethylzinc addition to benzaldehyde and the excellent catalytic activity was recorded when the reaction was carried out in toluene at rt using 5 mol% of catalyst and a low benzaldehyde/Et$_2$Zn ratio of 1/1.1 (>99% conversion and 93% ee). The best reaction conditions were used with other aromatic aldehydes and yields and ees were similar to those observed with benzaldehyde except for pyridine carboxaldehyde. In this case, nitrogen atom of pyridine ring was suspected to be too chelating. Excellent enantioselectivity (93%) was also obtained with an enolizable aldehyde, n-octanal. Furthermore the catalyst could be recovered by simple filtration and could be reused five times without loss of catalytic activity.

The results obtained are interesting since the supported-ligands containing a NH group gave generally poor to moderate enantioselectivities.

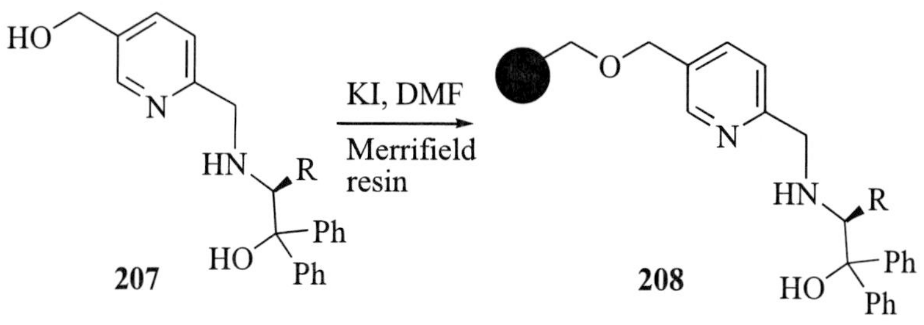

Scheme 85.

El-Shehawy studied the effect of the loading of the ephedrine moiety and the degree of crosslinking of polymer-supported ephedrine for the diethylzinc addition to benzaldimine [141].

Different chiral polymers were prepared by copolymerization of monomers 141 with styrene and DVB (Scheme 68). The compositions of these polymers are summarized in Table 14.

Table 14. Compositions of polymers 209-213

Chiral Polymer	Chiral Monomer	Molar Ratio[a]	Loading of ligand (mmol/g) (DF)[b]
209	(1S,2R)-141a	1:7:2	0.764 (0.10)
210a	(1S,2R)-141b	1:7:2	0.771 (0.10)
210b	(1S,2R)-141b	1:8:1	0.750 (0.10)
210c	(1S,2R)-141b	1:6:3	0.821 (0.10)
210d	(1S,2R)-141b	1:5:4	0.735 (0.10)
210e	(1S,2R)-141b	2:6:2	1.292 (0.20)
210f	(1S,2R)- 141b	3:5:2	1.863 (0.30)
210g	(1S,2R)-141b	4:4:2	2.127 (0.40)
210h	(1S,2R)-141b	5:3:2	2.449 (0.50)
211	(1S,2R)-141c	1:7:2	0.792 (0.10)
212	(1S,2R)-141d	1:7:2	0.721 (0.10)
213	(1R,2S)-141b	1:7:2	0.807 (0.10)

[a] Molar ratio of copolymerization: chiral monomer:styrene:DVB.
[b] Degree of functionalization.

These chiral polymers **209-213** were then tested in the asymmetric addition of diethyzinc to benzaldimines. First, the catalytic activity of the copolymers **209, 210a, 211** and **2213** having a molar ratio of copolymerization of 1/7/2 were evaluated and the best result was obtained with polymer **210a** (92% yield and 82% ee) when the reaction was carried out in toluene at − 10 °C. The enantioselectivity of the reaction with polymeric ligand **210a** was highly temperature-dependant and surprisingly a significant decrease of the chemical yield when the reaction was performed at 0°C or rt was observed.

The increase of the crosslinking reagent (DVB) involved a significant decrease of both yield and ee. The same effect was observed when the loading degree of the aminoalcohol unit increased.

The optimized procedure with chiral polymer **210a** was employed with other aromatic aldimines and the same range of enantioselectivity was recorded.

El-Shehawy has reported the synthesis of original polystyrene-supported dendritic chiral ephedrine [142]. Chain end-functionalized polystyrenes bearing a number of 2, 4, 8 or 16 BnBr residues at their chain-ends, **PS(BnBr)₂-PS(BnBr)₁₆** were first isolated (Scheme 86).

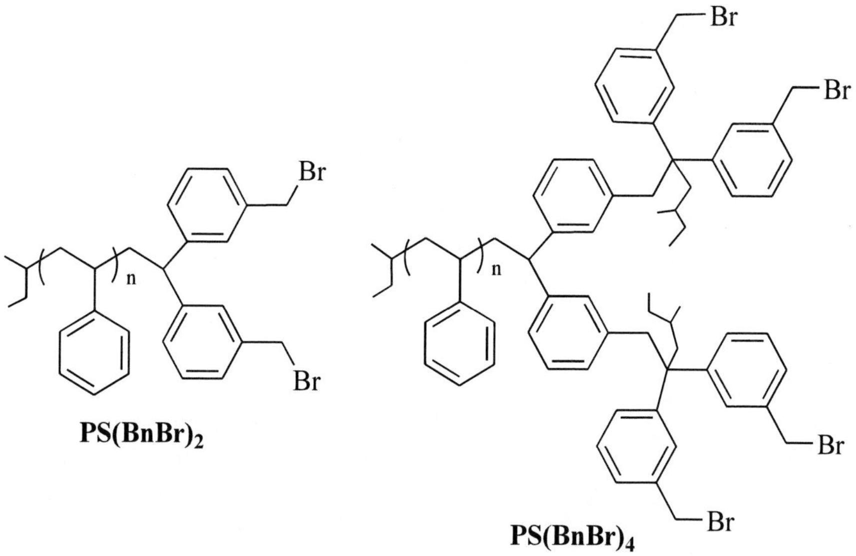

Scheme 86.

Reaction of polymers of **PS(BnBr)₂-PS(BnBr)₁₆** having two-sixteen BnBr moieties with ephedrine led to the corresponding polymers **PS(Ephed)₂-PS(Ephed)₁₆** bearing the same numbers of ephedrine at their chain-ends (Scheme 87).

The catalytic activity of these four chiral polymer-supported dendritic aminoalcohols was evaluated in the enantioselective diethylzinc addition to *N*-diphenylphosphinoylbenzaldimine. The reaction was first carried out in toluene at rt for 48 h using an equimolar amount of supported-ligand (based on the total numbers of ligand unit).

The first analysis of the results (Table 15) was that more chiral sites at chain-ends, higher the yield was except for **PS(Ephed)₁₆** (entries 1, 3, 5 and 6). But in that case, the problem is probably the proximity of the chiral sites and some site-site interactions could occur. For the four polymer-supported chiral ligands, the enantioselectivity was comprised between 79 and 90% which is relatively high. To obtain a satisfactory yield of 85-90% with

PS(Ephed)₂ and PS (Ephed)₄, a longer reaction time was necessary without any influence on the enantioselectivity (entries 2 and 4).

The best chiral polymer, **PS(Ephed)8a**, was also tested with other *N*-diphenylphosphinoylarylimine under the same reaction conditions and similar level of yield and enantioselectivity were observed.

The same trend as with imine was observed for the enantioselective diethylzinc addition to benzaldehyde and the best results (94% yield and 90 %ee) were also obtained with **PS(Ephed)₈** [143].

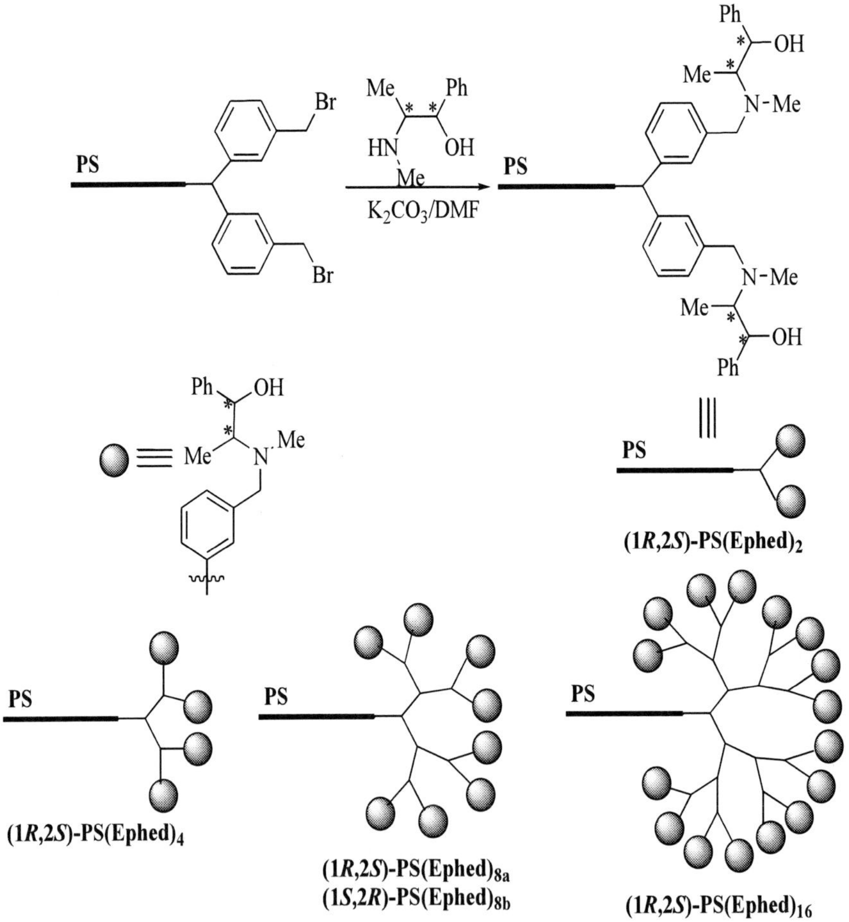

Scheme 87.

Although high catalytic activity was observed with polymer-supported aziridine-derived ligand **160**, the attachment on this type of chiral ligand through the phenyl group was studied [144]. The protected monomers **214** were prepared and copolymerized with styrene and DVB to afford the chiral aminoalcohol-containing copolymers **215** with a loading of 0.1 mmol/g (Scheme 88).

Table 15. Enantioselective diethylzinc addition to *N*-diphenylphosphinoylbenzaldimine using polymer-supported dendritic aminoalcohols

Entry	Supported-ligand	Reaction time (h)	Yield (%)	ee (%)
1	PS(Ephed)$_2$	48	53	81 (*R*)
2	PS(Ephed)$_2$	90	85	83 (*R*)
3	PS(Ephed)$_4$	48	71	84 (*R*)
4	PS(Ephed)$_4$	72	90	85 (*R*)
5	PS(Ephed)$_{8a}$	48	92[a,b]	90[a,b] (*R*)
6	PS(Ephed)$_{16}$	48	78	79 (*R*)
7[c]	PS(Ephed)$_{16}$	48	93	80 (*R*)
8	PS(Ephed)$_{8b}$	48	91	90 (*S*)

[a] 91% yield and 92% ee when using 1.0 molar of (1*R*,2*S*)-*N*-benzylephedrine as chiral ligand.

[b] 74% yield and 89% ee when using 1.0 molar of (1*R*,2*S*)-*N*-vinylbenzylephedrine copolymerized with styrene and DVB as chiral ligand.

[c] 1.5 molar equivalent of chiral polymer.

The three polymeric ligands were then examined in the diethylzinc addition to benzaldehyde. Protected polymer-supported ligand **215a** and **215b** were slighty less active than the Zwannenburg's polymeric ligand **160**. The trityl-protected supported-ligand **215a** was also recycled and showed a complete loss of catalytic activity (3%) and enantioselectivity (0%).

It is important to pointed out the high enantioselectivity (84% ee) obtained with the unprotected polymer-supported ligand **215c** while the conversion was moderate (52 %) after 24h of reaction.

Polymer-supported chiral Schiff-base amino alcohols have also been used as chiral ligands. The four polymeric Schiff-base ligands **216a**, **216b**, **217a** and **217b** have been anchored onto Merrifield resin (1 mmol Cl/g, 1% DVB, 200 mesh) using the corresponding chiral ligands (Scheme 89) [145]. These chiral polymers have been first tested as chiral ligands in the phenylacetylene addition to acetophenone and chiral polymer **217a** induced the best enantio-selectivity (63%) when the reaction was performed in toluene at rt using 10 mol% of the catalyst. When the reaction temperature decreased to – 18 °C, a significant increase of the enantioselectivity (75%) was observed. The optimized reaction conditions with chiral supported-ligand **217a** were then applied to different functionalized acetophenone and enantioselectivities were comprised between 62 and 89%. The polymeric ligand **217a** could be reused in five successive reactions with an average loss of enatioselectivity of 1% per run.

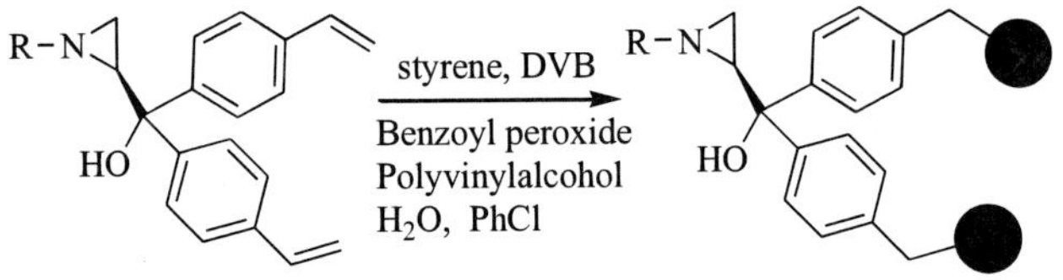

214a: R=CPh$_3$
214b: R=CHPh$_2$

215a: R=CPh$_3$
215b: R=CHPh$_2$
215c: R=H

Scheme 88.

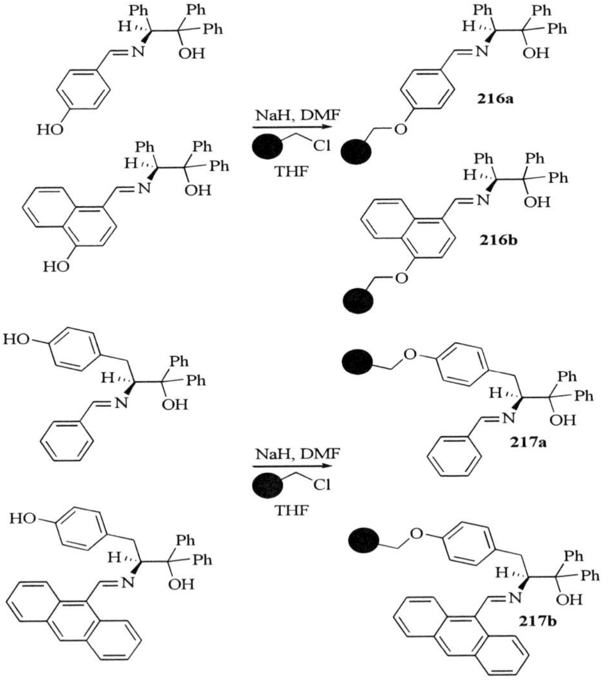

Scheme 89.

5.2. SUPPORTED SULFONAMIDE
AND *N*-SULFONYLATED AMINOALCOHOLS

Itsuno has described the first enantioselective allylation of aldehydes using crosslinked polymer-supported *N*-sulfonylated aminoalcohols as chiral ligands [146]. Polymer-supported ligands **220** and **221** were prepared by copolymerization of the norephedrine-derived monomer **218** or D-camphor-derived **219** with styrene and divinylbenzene (DVB) in a [chiral monomer]/[styrene]/[DVB] molar ratio of 1/8/1 (Scheme 90).

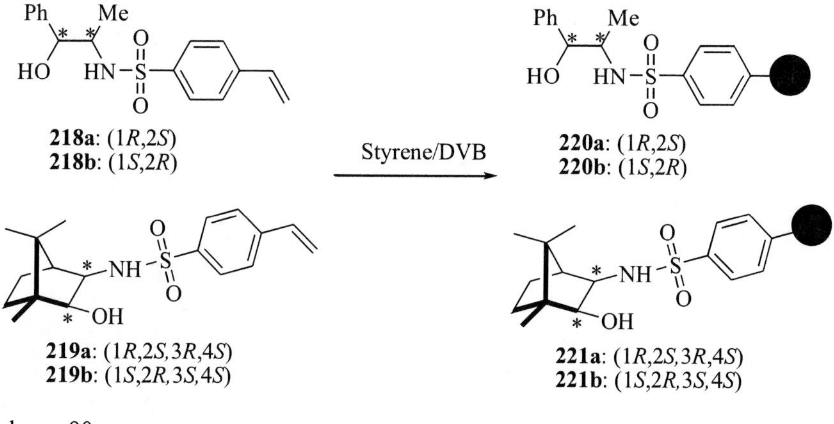

218a: (1*R*,2*S*)
218b: (1*S*,2*R*)

Styrene/DVB

220a: (1*R*,2*S*)
220b: (1*S*,2*R*)

219a: (1*R*,2*S*,3*R*,4*S*)
219b: (1*S*,2*R*,3*S*,4*S*)

221a: (1*R*,2*S*,3*R*,4*S*)
221b: (1*S*,2*R*,3*S*,4*S*)

Scheme 90.

Reaction of these chiral polymer-supported ligands with triallylborane led to the corresponding allyboron reagents which were allowed to react with aldehydes. When the norephedrine-derived ligands **220** were employed in the asymmetric allylation of benzaldehyde, homoallylic alcohol (*S*) was isolated in a 93% yield and in 75% ee. The enantioselectivity was slightly higher with the camphor-derived analog **221a** (85% ee). The polymer-supported ligands induced better enantioselectivities than the non-polymeric analogs. High enantioselectivities were also observed in the allylation of aliphatic aldehydes (92% for acetaldehyde and 84% for pivalaldehyde).

Liskamp has reported the synthesis and the screening in asymmetric catalysis of a library of polymer-supported peptidosulfonamide [147]. (*R,R*) or (*S,S*)-pyrrolydines **222** prepared in six steps from D- or L-tartaric acid were anchored onto Argonaut resin (0.41 mmol/g) to afford the corresponding supported pyrrolidines **223** (Scheme 91). The polymeric peptidosulfonamides **224 and 225** were then synthesized in four steps from the immobilized pyrrolidines **223**. The different chiral polymers were tested in the titanium-mediated diethylzinc addition to benzaldehyde, *p*-chlorobenzaldehyde, cyclohexanecarboxaldehyde and phenylacetaldehyde. Both yield and ee were very low with all these supported-ligands for the enantioselective reaction with the two aliphatic aldehydes. With aromatic aldehydes, the best results were observed with both leucine-derived supported peptidosulfonamides **224d** and **225d**.

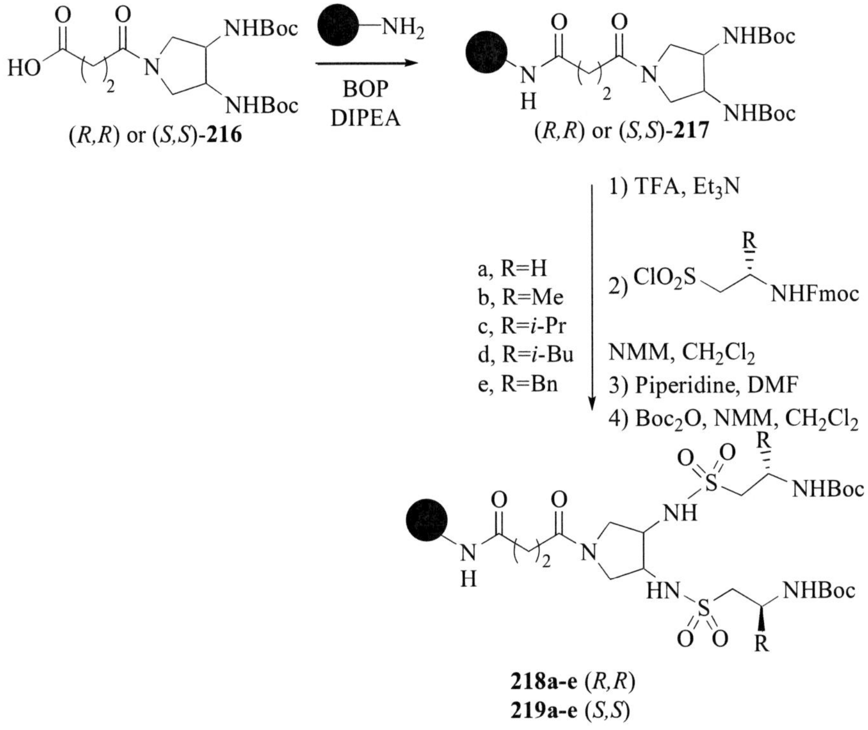

Scheme 91.

Isoborneolsulfonamide-containing *trans* 1,2-diaminocyclohexane have been immobilized on solid support [148]. The four polymer-supported ligands **227** were synthesized by suspension copolymerization in polyvinyl alcohol (average Mw=85,000-146,000) of chiral monomer **226** with styrene and DVB (Scheme 92). These four chiral polymers **227** were then employed as chiral ligands in the enantioselective titanium-mediated diethylzinc addition to acetophenone. Although enantioselectivities were excellent (99%) with these four polymer-supported ligands, the chemical yields were low (16 (**227a**), 23

(**227b**) and 36% (**227d**)) to moderate (56% (**227c**) even after a very long reaction time (17 days).

A more reactive zinc reagent, the ethyl phenyl zinc reagent prepared by transmetallation of diethylzinc with triphenylboron was employed. This reagent was used with *p*-bromoacetophenone in the presence of 5 mol% of polymeric ligand **227** and an excess of titanium tetraisopropoxide and the corresponding tertiary alcohol was isolated with good enantioselectivities (71-86%). The amount of the crosslinking agent had a strong influence on the catalytic activity and the enantioselectivity and the best results were obtained with the soluble polymer-supported ligand **227d**.

This ligand was employed in the enantioselective phenyl addition to different ketones, and the same level of chemical yield and enantioselectivity was observed for aromatic ketones while the enantioselectivity dropped significantly with hexan-2-one (38% ee).

The reuse of the polymer-supported ligand **227d** was studied and the catalytic activity decreased rapidly after only two runs.

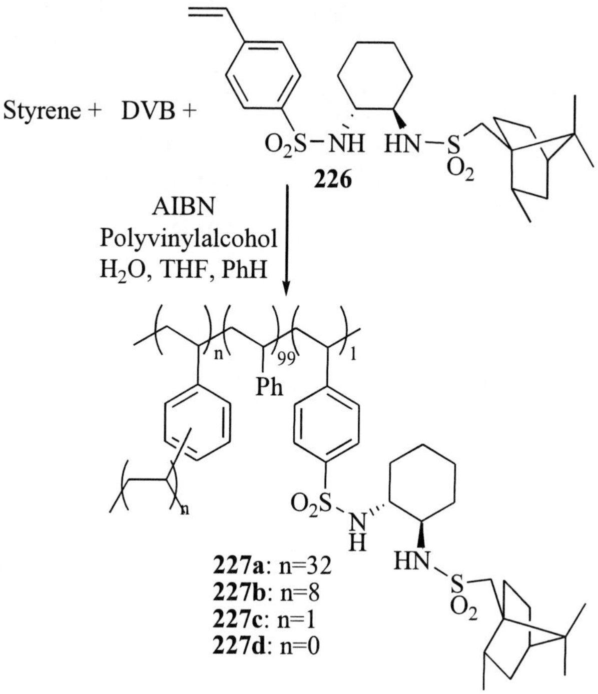

Scheme 92.

Gau has prepared three different types of polymer-supported *N*-sulfonylated aminoalcohols. The first one **230** was produced by anchoring the aminoalcohol **229** through the nitrogen atom to a chlorosulfonated resin (1 mmol Cl/g; 2% DVB) with a ligand loading of 2.30 mmol/g (Scheme 93) [149]. In order to decrease the ligand unit loading, the authors have also synthesized chiral monomer **231** which was allowed to copolymerize with DVB and styrene to lead to polymeric ligands **232a** and **232b** with a ligand loading of 1.28 mmol 1.76 mmol, respectively.

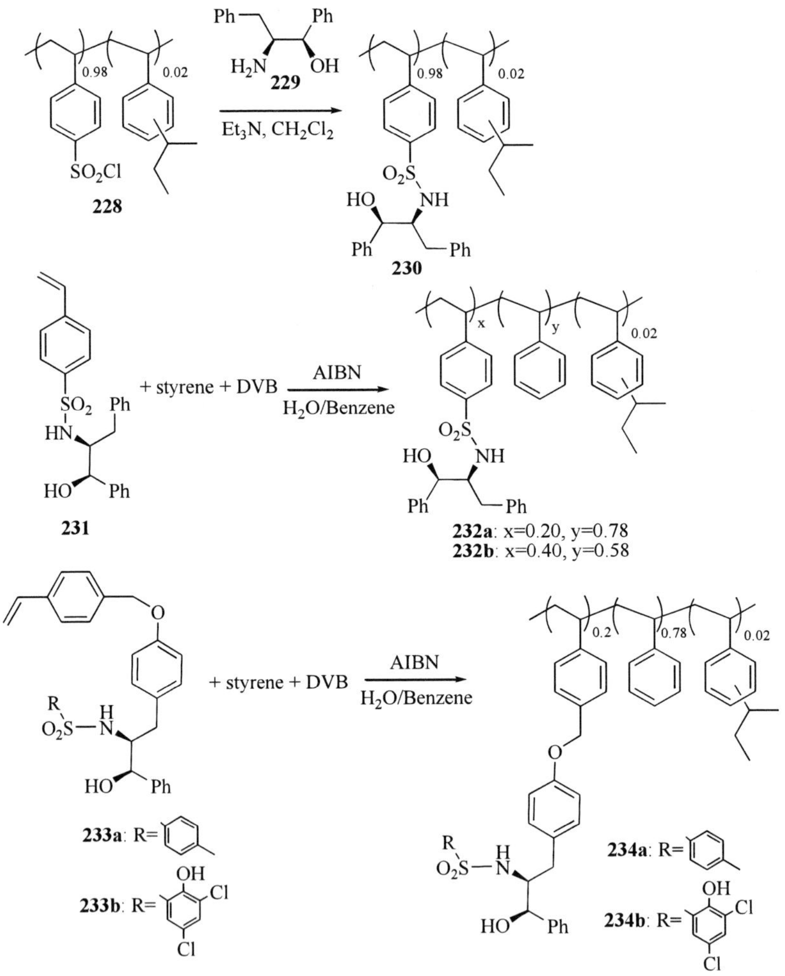

Scheme 93.

It was not surprising to observe moderate yield (62%) and enantio-selectivity (44%) for the titanium-mediated diethylzinc addition to benzaldehyde with the polymeric ligand **230** since a high ligand-loaded polymer is generally less active because of the proximity of the catalytic sites. The best result (100% yield and 92% ee) was obtained with 10 mol% of the low ligand-loaded **232a** in the presence of 10 equivalents of titanium isopropoxide (relative to the ligand). This polymer-supported ligand was also employed with different aldehydes and both yield an ee were high in all cases. This latter could be also reused five times but the enantioselectivity decreased from 92 to 86%.

In the aim to discover other polymer-supported catalysts having a better activity and reusability, chiral polymers **234a** and **234b** were prepared by copolymerization with styrene and DVB with respectively a ligand loading of 0.93 mmol/g and 0.95 mmol/g [150]. The heterogeneous **234a**/Ti(OiPr)$_4$ system proved to be better than the **232**/ Ti(OiPr)$_4$ systems for the enantioselective diethylzinc to aldehydes and a highly recoverable catalyst. Indeed, this polymeric ligand/Ti(OiPr)$_4$ catalyst could be reused 9 times with enantioselectivities $\geq$ 87% ee.

The polymer-supported tridentate ligand **234b** was employed in the enantioselective triethylaluminium addition to aldehydes. To obtain the same level of enantioselectivity as precedent system, 28 equivalents of Ti(OiPr)$_4$ relative to the chiral ligand was necessary. With most aldehydes used, the enantioselectivities were slightly lower than those obtained with the heterogeneous **234a**/Ti(OiPr)$_4$ system.

5.3. SUPPORTED OXAZOLINE OR OXAZOLIDINE AS LIGANDS

Two polymer-supported chiral aminooxazoline **236** have been prepared because of better chelating properties of this type of ligand due to the higher nucleophilicity of the amino group [151]. Polymers **236** were produced after reaction of polymer-supported isatoic anhydride **235** (isolated from chloromethylated styrene-DVB polymer (4.8 eq. Cl/g)) and the corresponding aminoalcohol under acidic conditions (Scheme 94).

The catalytic activity of these polymeric oxazolines was examined in the enantioselective diethylzinc addition to benzaldehyde. Chiral supported-oxazoline **236b** exhibited slightly better enantioselectivity (89%) and yield

(90%) than its analog **236a** when the reaction was carried out in a 20% CH$_2$Cl$_2$/toluene mixture at 0°C. This chiral supported-ligand has been recycled three times without a significant loss of performance.

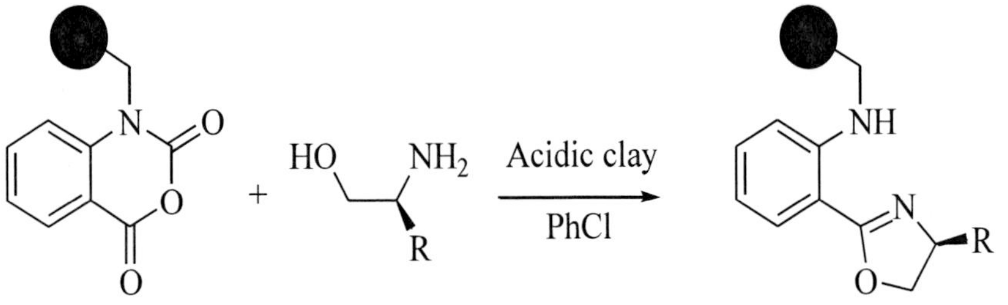

235

236a: R=Bn
236b: R=i-Bu

Scheme 94.

Chiral ferrocenyl oxazoline-based ligand which is often used in asymmetric catalysis have been immobilized onto two different supports, an insoluble trityl chloride polystyrene and a soluble polymer, a polyethylene glycol monomethyl ether [152]. The linker-containing compound **237** has been anchored on both polymers to produce respectively chiral insoluble polymer **238** and chiral soluble polymer **239** (Scheme 95). These polymer-supported ligands were then tested in the enantioselective phenyl and ethyl addition to *p*-chlorobenzaldehyde and benzaldehyde respectively. Polystyrene-supported oxazoline **238** was totally inefficient for the asymmetric phenyl transfer to *p*-chlorobenzaldehyde while a high enantioselectivity (87%) was observed for the ethyl addition to benzaldehyde. Contrary to the insoluble one, the soluble polymer **239** exhibited high enantioselectivity for both reactions (97 and 86% respectively). In general, the enantioselectivities were slightly lower than those obtained with the non-polymeric ligand. Futhermore, no significant loss of enantioselectivity was detected after five consecutive phenyl additions to *p*-chlorobenzaldehyde with the recovered polymer.

Portnoy focused his attention on the immobilization of PyBox ligands [153]. The polymer-supported PyBox **240** were prepared in five steps from the Wang trichloroacetamidate resin (Scheme 96). The supported-ligands **240** were allowed to react with copper (I) triflate for 24h to afford the corresponding catalysts **241**. These catalysts were then used in the first heterogeneous catalyzed addition of phenylacetylene to imine.

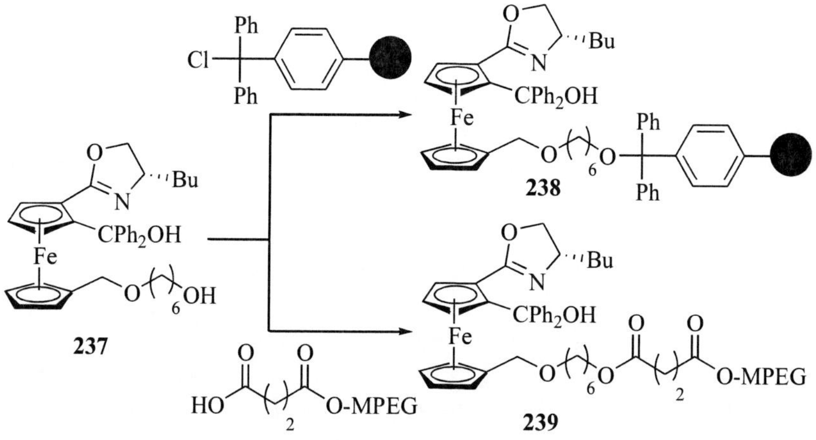

Scheme 95.

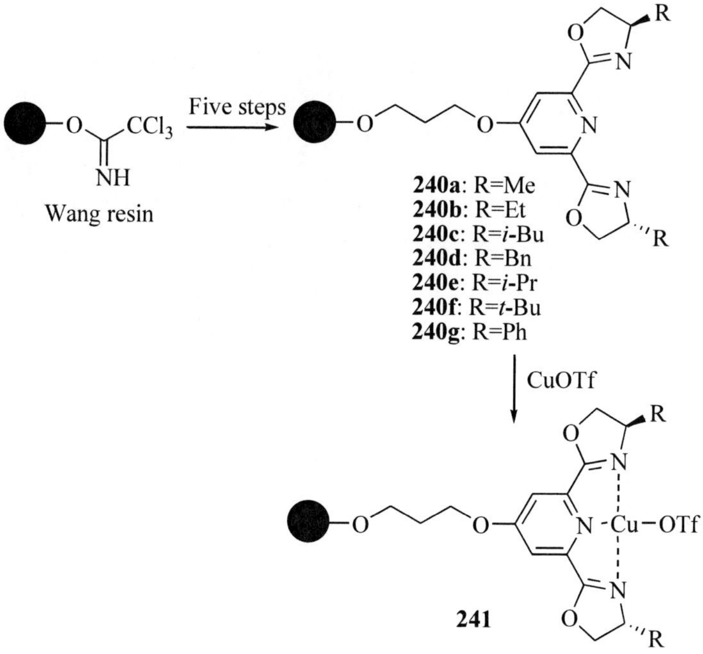

240a: R=Me
240b: R=Et
240c: R=*i*-Bu
240d: R=Bn
240e: R=*i*-Pr
240f: R=*t*-Bu
240g: R=Ph

Scheme 96.

The best enantioselectivity (83%) was obtained when the reaction was performed in dichloromethane at 40°C for 24h using 10 mol% of the catalyst **241f** which possesses a steric group on the oxazoline ring. When a recovered catalyst was employed for a second run, a drop of the catalytic was observed which is probably due to the oxidation of the metal. The use of other solvents such as THF in order to improve the recyclability was attempted but a decrease of the catalytic activity (80 to 56%) was detected during three consecutive reactions while the enantioselectivity increased (54 to 60%). To reduce the oxidation, ascorbic acid was added to the reaction mixture and proved to preserve the catalytic activity but the enantioselectivity was totally lost.

Another strategy to immobilize a PyBox ligand has been described by Moberg and Levacher [154]. Supported PyBox 243 has been prepared using click-chemistry reaction between alkyne 241 and polystyrene-supported azide (1.2 mmol N₃/g) (Scheme 97). The PyBox loading was estimated at 0.8 mmol/g. This polymer-supported PyBox/Cu(I) catalyst was then tested in the phenylacetylene to N-benzylideneaniline. The enantioselectivity was lower than that observed with the non-polymeric but slightly better than those with Portnoy's immobilized PyBox 240. Compared to catalyst 241 similar trend was observed concerning the recyclability. When the immobilized PyBox-Cu(I) catalyst was only filtered, a drop of both catalytic activity and enantioselectivity were detected after five consecutive runs. When the supported-ligand was washed several times and reloaded with copper triflate, catalytic activity and enantioselectivity were recovered. The same level of conversion and ee were observed with other aromatic imines.

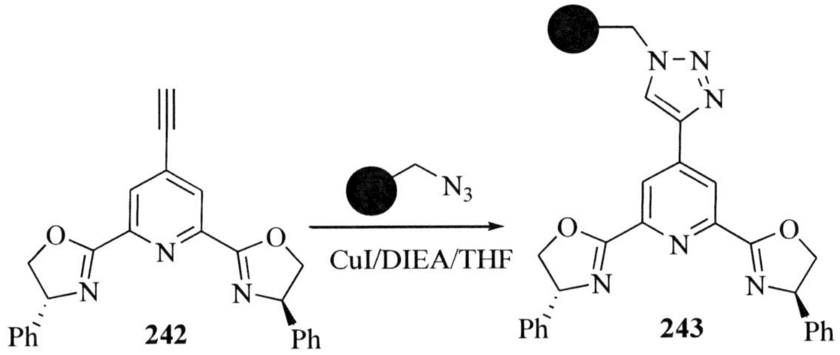

Scheme 97.

The use of polymer-supported oxazolidine derived from (1R,2S)-cis-1-amino-2-indanol for the asymmetric alkynylation of aldehyde has been reported [155]. Three chiral supported-oxazolidine 244 were prepared in two steps from Merrifield resin (2.5 mmol Cl/g) (Scheme 98).

The optimization of the enantioselective addition of phenylacetylene to benzaldehyde has been done with the polymeric ligand 244a. The best result

(82% yield and 79% ee) was obtained when the reaction was performed in THF at room temperature for 24 h in the presence of Et$_2$Zn, Ti(OiPr)$_4$ and 10 mol% of the ligand. The two other polymeric ligands **244b** and **244c** induced higher enantioselectivity (83% and 90% respectively) under the optimized reaction conditions. The same level of catalytic activity and enantioselectivity was observed with a variety of aromatic and heteroaromatic aldehydes when the polymer-supported oxazolidine **244c** was employed. This supported ligand could be reused three times without significant loss of performance.

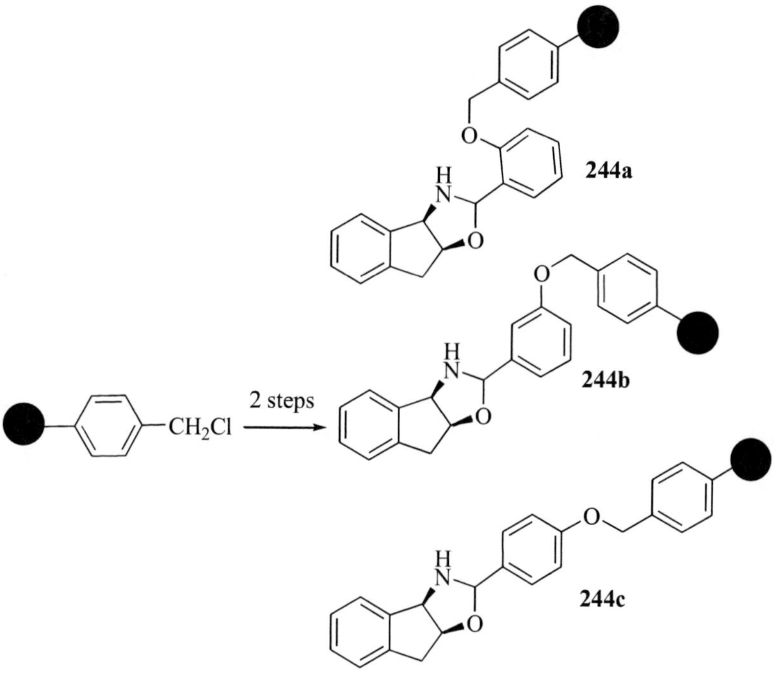

Scheme 98.

5.4. SUPPORTED SALEN AS CHIRAL LIGANDS

The catalytic activity of polymer-supported chiral salen has been also studied in the enantioselective diethylzinc addition to aldehydes since the importance of this type of ligand in asymmetric catalysis. Soluble polymer-supported salens have been synthesized by Venkataraman [156]. Salen **245** was prepared and was then anchored on MeO-PEG changing only the spacer between the PEG matrix and the salen to lead to the polymer-supported salens **246-248** (Scheme 99).

The enantioselective diethylzinc addition to benzaldehyde was then carried out using 10 mol% of polymeric salens **246-248** and the best result (90% yield and 82% ee) was obtained when the chiral polymer was used. With other functionalized aromatic aldehydes, both yield and ee were lower. Futhermore the soluble polymers could be recovered and reused whithout significant loss in selectivity or reactivity.

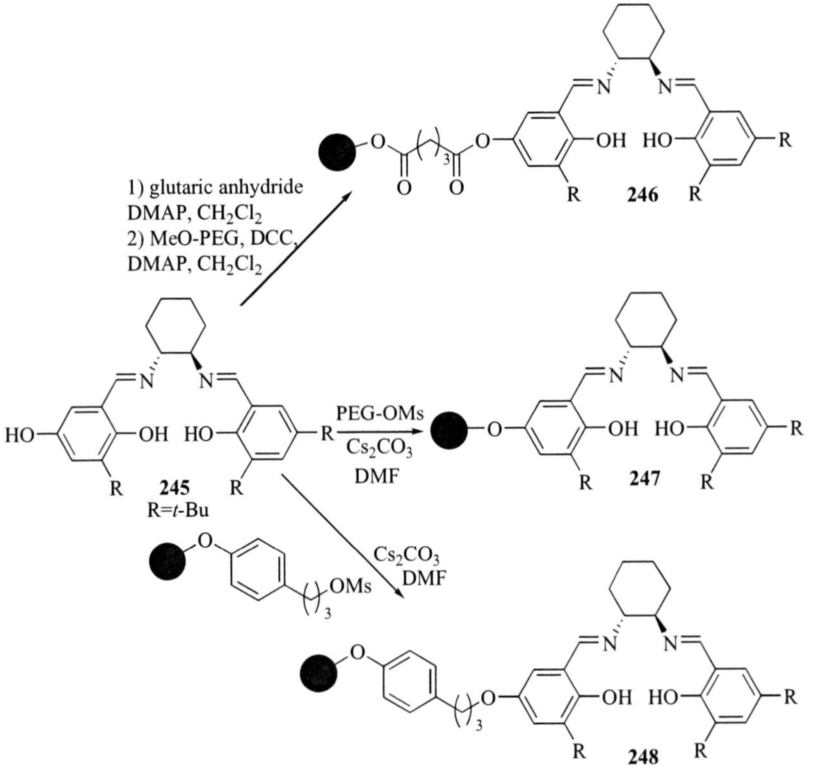

Scheme 99.

Chiral salen-type linear polymers have been prepared and tested in the diethylzinc addition to aldehydes [157]. The symetrical dialdehyde-containing compound **249** was first synthesized in five steps from 2-*tert*-butylphenol and allowed to react with 1,2-diaminocyclohexane and DPEN to afford the linear polymers **250** (Scheme 100). The catalytic activity of these chiral linear polymers was evaluated in the diethylzinc addition to *m*-nitrobenzaldehyde using 1 mol% of **250** (with respect to the monomeric unit). Although yields were satisfactory the ee were moderate and chiral linear polymer **250b** proved to be more efficient for the reaction. The recovered polymer has been reused without loss of performance.

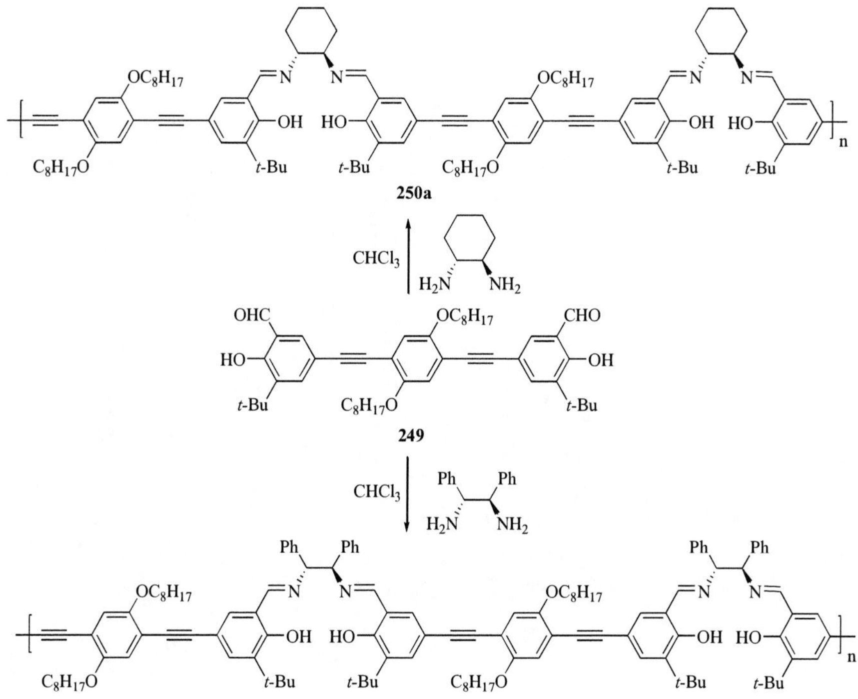

Scheme 100.

Chapter 6

ASYMMETRIC π-ALLYLIC SUBSTITUTION

Another largely studied asymmetric reaction to form new carbon-carbon bond is the palladium-catalyzed allylic substitution using particularly homogenous ligand (Scheme 101) [158]. Among the ligands employed, numerous phosphine-containing ones have been reported and we report here only the phosphorus-free nitrogen-containing ligands. The use of hetero-geneous ligands has not been studied extensively and the use of some immobilized nitrogen-containing ligands has been published.

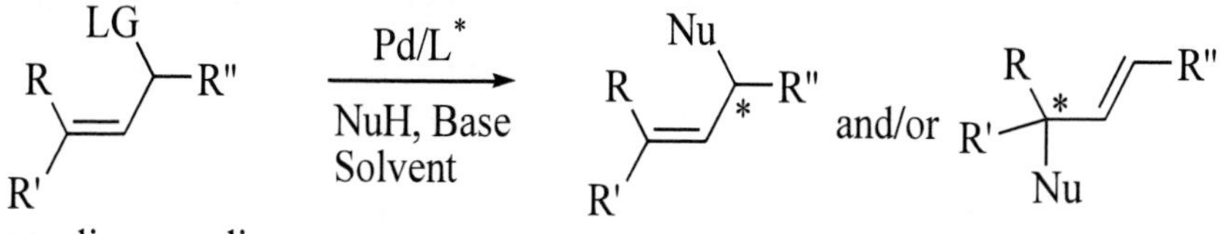

acyclic or cyclic
LG: Leaving group

Scheme 101.

To the best of our knowledge, the first polymeric ligand for this reaction was those of Lemaire's group [20]. Their chiral polyamide **23** and poly(urea) **24** (Scheme 13) was employed for the catalytic allylic alkylation. Using **23** the conversion was low (38%) and the ee reached 80% while with **24**, the conversion was better (72%) and enantioselectivity was lower (38%). In both cases, conversion and ee were lower than those obtained with the corres-ponding monomeric diamine.

Moberg focused her attention on the immobilzation of oxazoline-derived ligands for the enantioselective allylic alkylation [159]. Pyridine-containing oxazolines **251** were first prepared and were supported on TentaGel resin to afford the corresponding polymeric oxazolines **252** (Scheme 102). These polymer-supported ligands were then tested in the palladium-catalyzed allylic substitution and the sterically bulky ones **252b** and **252c** proved not to be active (< 5% yield).

When the chiral polymer **252a** was used as chiral ligand, the product was isolated in a yield comprised between 60 and 100% and in 80% ee; this result was closed to that obtained with the monomeric ligand.

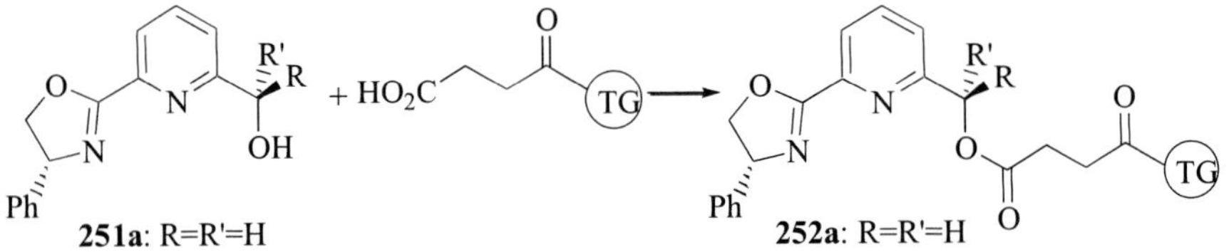

251a: R=R'=H
251b: R=*t*-Bu, R'=H
251c: R=H, R'=*t*-Bu

252a: R=R'=H
252b: R=*t*-Bu, R'=H
252c: R=H, R'=*t*-Bu

Scheme 102.

Polymer-supported bis(oxazoline) **109** having a degree of functionalization of 0.071 mmol/g have been tested in the same reaction [102]. This polymeric ligand exhibited better enantioselectivity (94-95%) than the previous chiral polymers **252**.

The asymmetric π-allylic alkylation using a polymer-supported pyridyl-diamide as chiral ligand was applied to the synthesis of the (*R*)-baclofen which is an agonist of the GABA$_B$ receptor [160]. The unsymmetrical cyclohex-anediamine **253** was anchored onto a TentaGel HL-COOH resin using a classical coupling reaction to give the polymer-supported ligand **254** (Scheme 103). This one was then tested in the molybdenum-catalyzed allylation reaction of dimethylmalonate with 3-phenylprop-2-enyl methyl carbonate. A high branched-to-linear ratio (35/1) and enantioselectivity were obtained with the polymeric ligand and this result was quite similar to those observed with the non-supported ligands. The ligand could be recovered by filtration and could be reused for seven consecutive cycles without loss of catalytic activity and enantioselectivity.

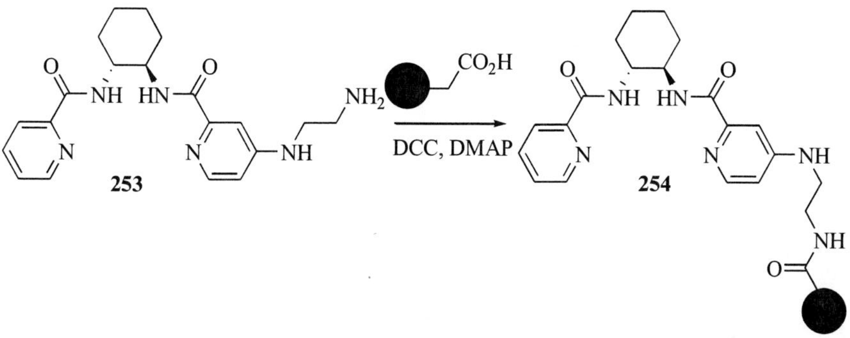

253

254

Scheme 103.

Bandini has described the grafting of diamino-oligothiophene compounds onto the monomethyl ether of PEG5000 to give the soluble polymer **256** (Scheme 104) [161]. High yield (98%) and ee (99%) were observed with soluble polymer **256a** when the allylic alkylation reaction between 1,3-diphenylallyl and dimethyl malonate was performed in THF at room temperature using cesium carbonate as the base. This result is comparable to that obtained with the non-polymeric ligand. Polymer-supported ligand **256b** proved to be less efficient than **256a**. The recyclability of the supported-ligand/Pd catalyst was also evaluated and loss of performance was detected only after three cycles.

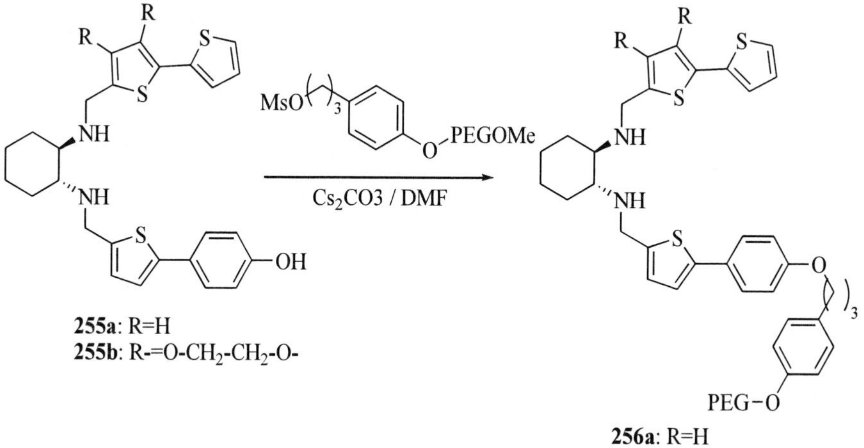

255a: R=H
255b: R-=O-CH_2-CH_2-O-

256a: R=H
256b: R-=O-CH_2-CH_2-O-

Scheme 104..

Chapter 7

DIHYDROXYLATION
AND AMINOHYDROXYLATION

The osmium-catalyzed asymmetric dihydroxylation (AD) is an excellent method to prepare chiral vicinal diols [162]. Most of the ligands employed are cinchona alkaloids and are derived from quinine or quinidine like ligand **257** for example. In 1992, Sharpless has described the improvement of the osmium-catalyzed asymmetric dihydroxylation by using bis cinchona alkaloids such as 1,4-bis(9-*0*-dihydroquinyl)phthalazine (**258**) as chiral ligand (Scheme 105) [163]. This type of ligand has then received much attention since high enantioselectivities were observed. Although the amount of OsO$_4$ has been reduced with the possibility to use cooxidants, the high cost of osmium and chiral ligands, the toxicity, the possible contamination of the chiral products by osmium avoid the use of this process in industry. The possibility to support and to recycle the osmium/ligand catalyst and particularly the use of soluble or insoluble supported ligands has been explored. In order to recycle the catalyst, number of research groups has investigated the immobilization on soluble and insoluble polymers of this type of alkaloids

7.1. INSOLUBLE SUPPORTED ALKALOID AS LIGAND

The first work about AD using heterogeneous ligand has been reported by Sharpless [164]. The polymeric alkaloids **259-262** (Scheme 106) were prepared by copolymerization of the corresponding monomer with acrylonitrile in the presence of the radical initiator AIBN. The performance of these polymer-supported cinchona were evaluated in the asymmetric dihydroxylation of *trans* stilbene using OsO$_4$ and NMO or K$_3$Fe(CN)$_6$ as secondary oxidant. The reaction rate was slow and no ee was detected using the polymeric ligand **259** probably because of the proximity of the active site to the polymer matrix. The best results (81-87% yield and 85-93% ee) were observed when the supported alkaloid **260** was used as chiral ligand in the presence of only 1 mol% of OsO$_4$ and NMO in acetone/water at 10°C. It is interesting to note that when the OsO$_4$/supported ligand complex was reused a second time without further addition of OsO$_4$, the yield and enantioselectivity dropped slightly. Both yield and ee were improved when K$_3$Fe(CN)$_6$ was used as cooxidant with polymers **261** and **262**.

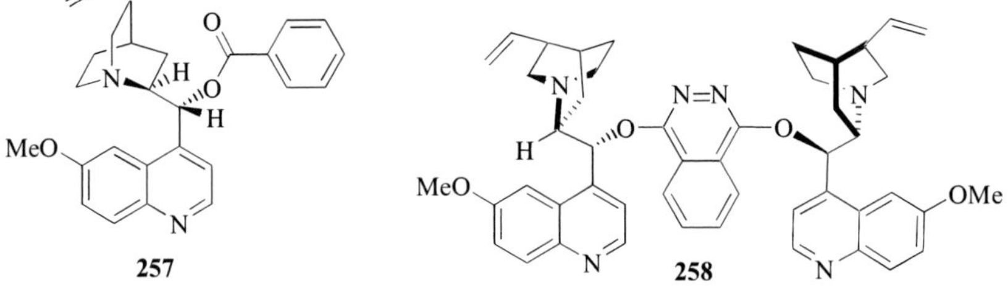

257   258

Scheme 105.

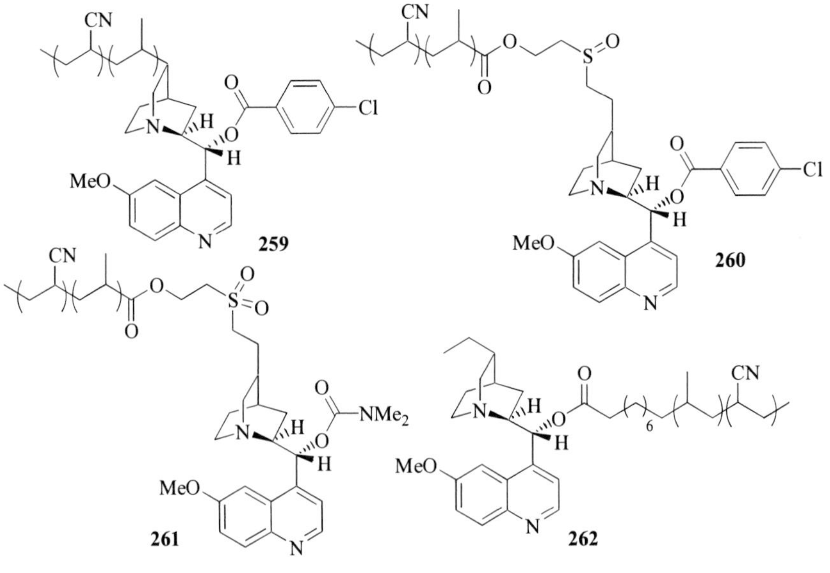

259   260

261   262

Scheme 106.

When the same reaction was tested with different olefins using the acetate analog of polymer **259**, the polymer-supported alkaloid **263**, low enantioselectivities were observed (Scheme 107) [165]. High chemical yields could be obtained when low alkaloid-loaded copolymers were used (< 15 mol%). Moreover high degree of alkaloid incorporation seemed to inhibit the reaction. Use of chiral polymers **264** and **265** did not improve the enantioselectivity of the dihydroxylation of *trans* stilbenes and ee were very low with polymer **264** compared to those obained with polymers **259-260**.

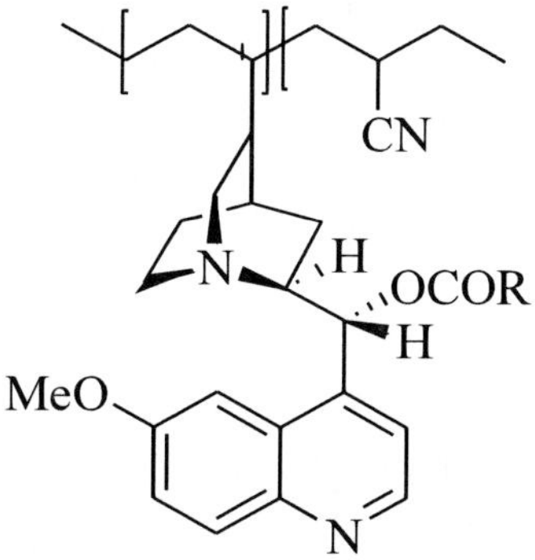

263: R=CH$_3$
264: R=p-Cl-C$_6$H$_4$
265: R=m-(CH$_3$O)-C$_6$H$_3$

Scheme 107.

Dihydroquinidine **266** containing a *p*-vinylbenzotate was prepared and was copolymerized either with styrene or with 4-phenylstyrene (Scheme 108) [166]. The effect of the loading of the alkaloid monomer on the activity of the corresponding polymer for the AD of *trans* stilbenes was examined and the results are summarized in Table 16. As already observed with different supported ligands, higher the loading of the alkaloid monomer, more important drop of chemical yield, reaction rate and enantioselectivity were. Polymer **267a** proved to be more efficient than its homolog **267b** and the complex **267a**/OsO$_4$ could also be reused without significant loss of performance. This system was employed for the asymmetric dihydroxylation of other olefins including non-aromatic and α,β-unsturated alkenes and moderate enantio-selectivities were obtained while chemical yields were high.

Table 16. AD of *trans* stilbene using polymers 267

Entry	Polymers 267		Reaction Time	Yield (%)	ee (%)
	n	m			
1	9.0	1.0	24 h	86 (68)[a]	82 (71)[a]
2	9.0	1.0	24 h	79[b]	78[b]
3	8.0	2.0	24 h	72 (65)	78 (68)
4	5.0	5.0	2 d	50 (50)	45 (40)
5	0.0	1.0	2-3 d	20	-

[a] Under brackets, yields and ee obtained with **267b**.
[b] Polymer containing dihydroquinine was used.

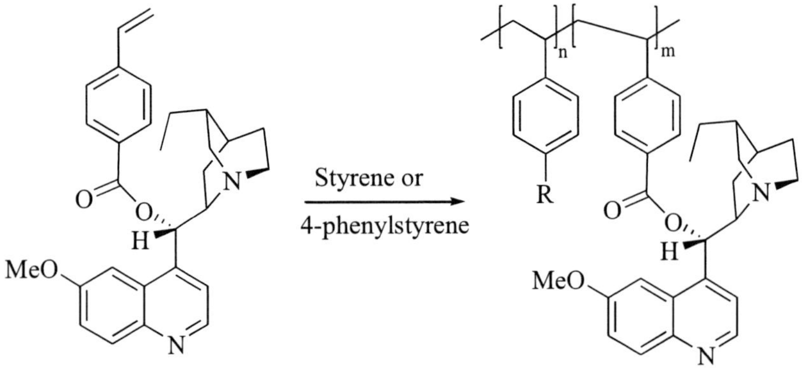

Scheme 108.

In order to improve the enantioselectivity obtained with their supported alkaloid for the AD, Salvadori has prepared polymer **269** which contained a spacer between the alkaloid and the polymer backbone [167]. This one was isolated after copolymerization of quinine-based monomer **268** with styrene and DVB in a 1/7/2 molar ratio (Scheme 109). This polymer-supported quinine was then tested in the AD of *trans* stilbene, styrene and (*E*)-β-methylstyrene. The enantioselectivities were higher than those obtained with polymers **263-264** with a shorter reaction time at rt (4-6h) or 0°C (7h). At 0°C, the enantioselectivity reached 87% for the dihydroxylation of *trans* stilbene. This polymer was also reused after recovering without significant loss of the activity.

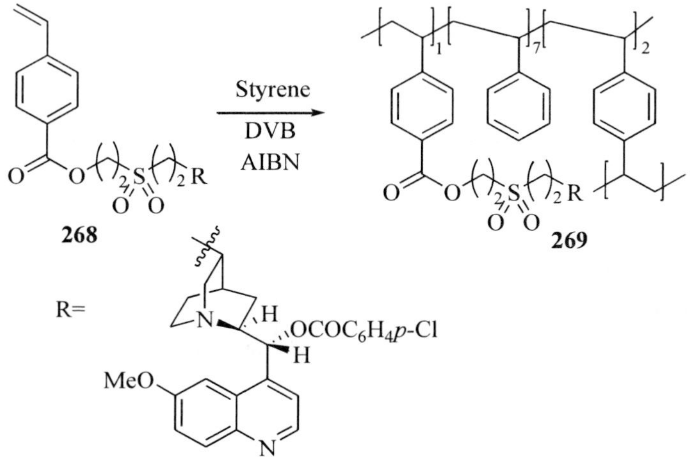

Scheme 109.

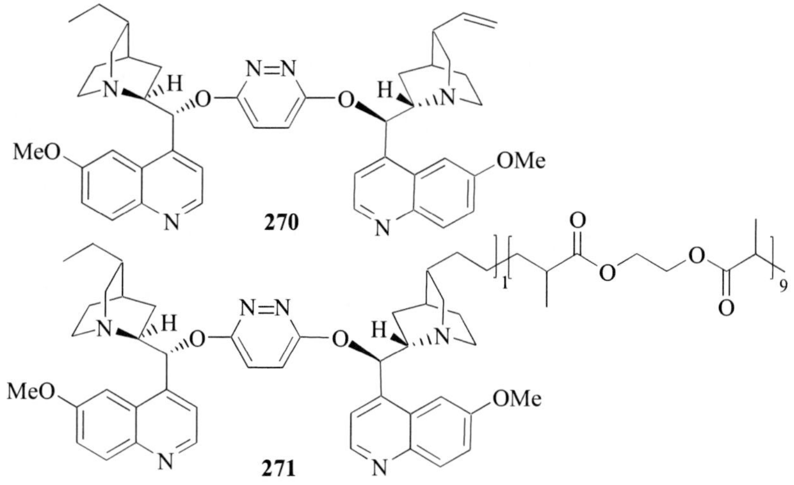

Scheme 110.

Lohray has reported the first example of immobilized bisdihydro-quinylpyridazine [168]. The monomer **270** has been prepared and was allowed to react with ethylene glycol methacrylate in a 1/9 ratio to afford the chiral polymer **271** (Scheme 110). When *trans* stilbene and styrenic olefins were dihydroxylated using supported ligand **271**, OsO$_4$ as oxidant and K$_3$Fe(CN)$_6$-K$_2$CO$_3$ as cooxidant, both yields and ee were high, ees being slightly lower than those observed with the homogeneous ligand. In opposite, the enantio-selectivities dropped significantly with aliphatic alkenes. The excellent results could be due to larger pore size and hydrophily of this type of polymer which facilitate the approach of the reactants to the catalytic site.

Since excellent results were obtained in the asymmetric amino-hydroxylation in homogeneous phase by Sharpless [169], heterogeneous systems appeared to be of great interest. Nandanan has reported the first hetero-geneous osmium tetroxide-catalyzed asymmetric aminohydroxylation of various olefins using polymer-supported bisdehydroquinine ligand **273** (Scheme 111) [170]. When chloramine T was used as nitogen source, yields and ee were moderate with all olefins.

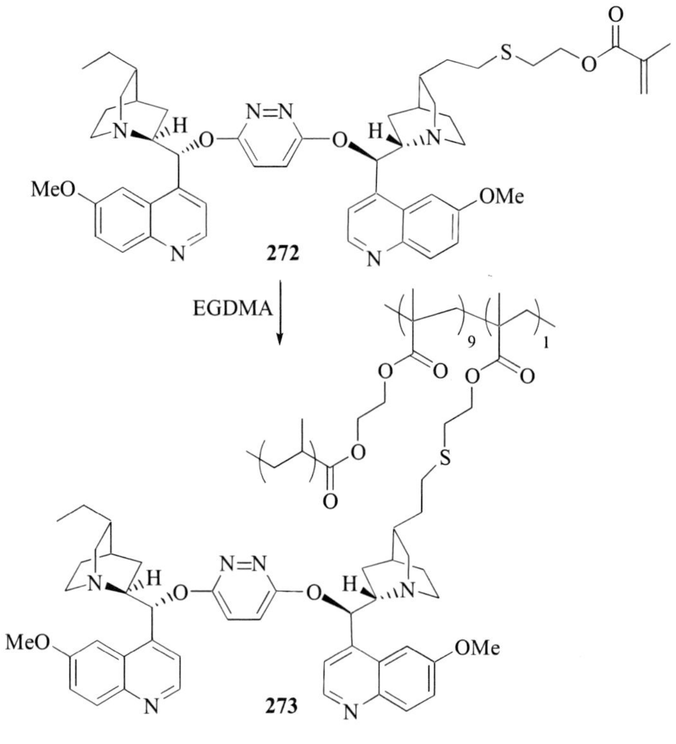

Scheme 111.

Monomer **268** was also copolymerized with hydroxyethyl methacrylate and ethylene glycol dimethacrylate as crosslinking agent in a 1/7/2 ratio to lead to the new polymer-supported alkaloid **274** (Scheme 112) [171]. This one was insoluble in all solvents and swelled very well in protic polar solvents. The catalytic activity of chiral polymer **274** was tested in the AD of various aliphatic or aromatic olefins using two different reaction conditions (NMO or $K_3Fe(CN)_6$ as cooxidant in respectively acetone/H_2O or t-BuOH/H_2O). With all alkenes used, the enantioselectivities were higher when $K_3Fe(CN)_6$ was employed as secondary oxidant and comparable to those obtained with the non-polymeric alkaloid. Concerning the aliphatic olefins, the enantioselectivities remained low to moderate.

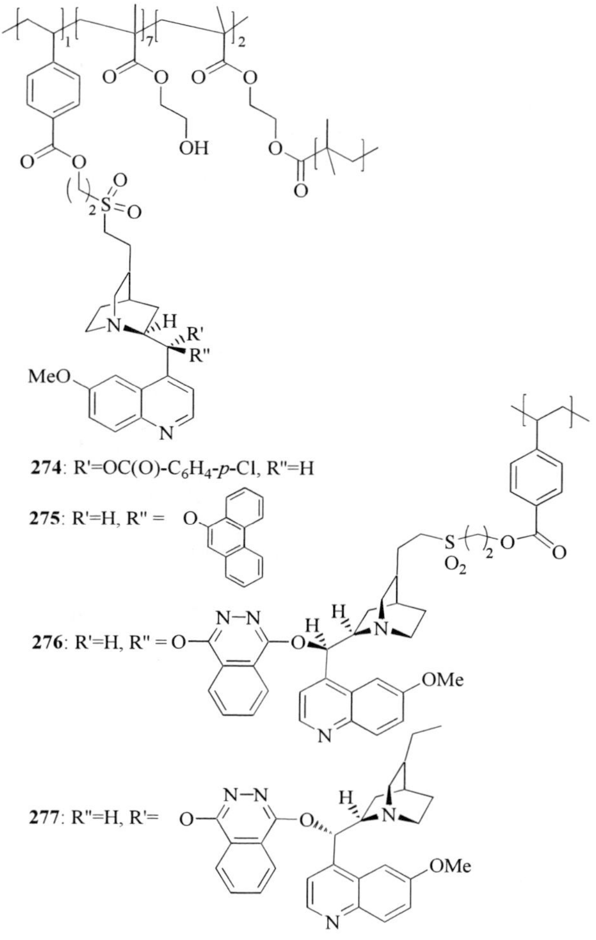

274: R'=OC(O)-C$_6$H$_4$-p-Cl, R''=H

275: R'=H, R'' =

276: R'=H, R'' =

277: R''=H, R'=

Scheme 112.

In order to improve the enantioselectivity in the AD of aliphatic alkenes, the two dihydroquinidine-derived polymers **275** and **276** were prepared by radical copolymerization of the corresponding monomers with hydroxyethyl methacrylate and ethylene glycol dimethacrylate as crosslinking agent

(Scheme 112) [172]. These two polymers **275** and **276** exhibited enhanced enantioselectivities (88 and 86% *vs* 75% with polymer **274**) for the dihydroxylation of two aliphatic alkenes, 1-decene and 5-decene compared to those observed with **274**. Furthermore, when polymer **276** was used as chiral ligand in the dihydroxylation of aromatic alkenes, ee were higher than 91% in all cases and close to those observed with the non-polymeric analog [173].

Salvadori has employed the polymer-supported bis(quinidyl)phthalazine **277** in the asymmetric aminohydroxylation of isopropyl cinnamate **278** in the presence of *N*-chloromethanesulfonamide sodium salt and K_2OsO_4 in *n*-propanol/H_2O (Scheme 113) [174]. The chemoselectivity, regioselectivity and ee are summarized in Table 17.

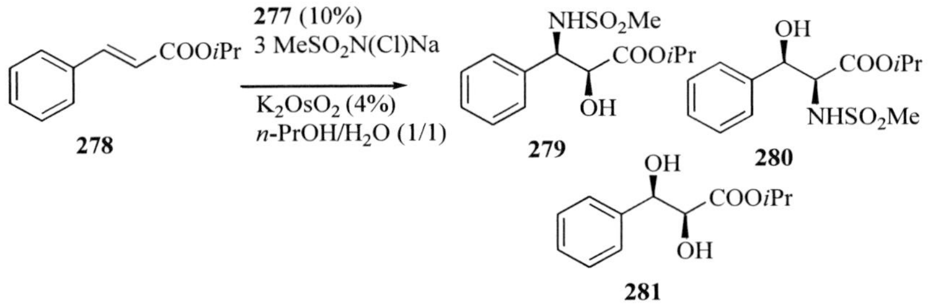

Scheme 113.

Table 17. Asymmetric aminohydroxylation with polymer 277

Entry	Conversion[a] (%)	Chemoselectivity[b] (%)	Regioselectivity[c] (%)	Ee of **279** (%)
1	98	94	91	87
2[d]	73	91	89	83
3[d]	58	84	89	81
4[e]	59	79	88	78

[a]**279+280+281**.
[b]**(279+280)/(279+280+281)**.
[c]**279/(279+280)**.
[d]Recycling run without addition of K_2OsO_4.
[e]Recycling run with addition of 10 mol% of K_2OsO_4.

Polymer **277** proved to be as efficient as homogeneous ligand in terms of activity, chemoselectivity and regio-selectivity. Furthermore, the enantio-selectivity was higher (87%) than those obtained with **273**.

When the osmium/ligand complex was recycled, both activity and chemoselectivy dropped significantly while chemoselectivity was remained. The enantioselectivity decreased slightly even when an initial amount of K_2OsO_4 was added to the mixture.

The polymer backbone polarity could influence the outcome of the reaction since the catalytic activity is strongly dependent on the solvent system. For this reason, the catalytic activity of homopolymers **284** and **285** and the copolymers **286** and **287** having a polar group on polymer backbone were compared [175]. Homopolymers **284**, **285a** and **285b** were prepared by polymerization of dihydroquinidine acrylate **282a**, dihydroquinidine 4-vinylbenzoate **283a** and dihydroquinine 4-vinylbenzoate **283b** respectively using AIBN as radical initiator (Scheme 114). These supported cinchona were tested in AD of *trans* methyl cinnamate and *trans* stilbenes using either NMO in acetone/H_2O or $K_3Fe(CN)_6$ in *t*-BuOH/H_2O. Analysis of the results showed that better enantioselectivities were observed with alkaloid 4-vinylbenzoate-derived homopolymers **285**. When the $K_3Fe(CN)_6$/*t*-BuOH/H_2O system was used with homopolymers, enantioselectivities were excellent with both alkenes.

Song assumed that in the acetone/H_2O solvent system, the homopolymers **285** formed a viscous lump which did not allow a high concentration of the substrate near the active site while in the *t*-BuOH/H_2O system these polymers swelled well. The accessibility of the active site is highly dependent on the compatibility between the polymer support and the liquid reaction medium. For this reason, copolymers **286** and **287** having a polar polymer structure were prepared by radical copolymerization of monomer **283** with methylmethacrylate or 2-hydroxyethyl methacrylate in the presence of AIBN (Scheme 114). Indeed, these polymers seemed to be more compatible with the liquid reaction medium because both yield and enantioselectivity were higher with the two acetone/H_2O and $K_3Fe(CN)_6$/*t*-BuOH/H_2O systems in the AD of *trans* methyl cinnamate.

The polymer **286** and **287**/OsO_4 complexes could be easily filtered from the reaction mixture and could be reused without significant loss of performance.

Since 1,4-bis(9-O-dihydroquininyl)phthalazine ((DHQ)$_2$-PHAL) **258** gave better results than the benzoate analog **257** (Scheme 105) in homogeneous catalysis, the heterogeneous ones were designed in order to check if the same trend was observed [176]. The copolymers **289a** and **289b** were prepared by copolymerization of the monomer **288** with methylmethacrylate or 2-hydroxyethyl methacrylate respectively (Scheme 115).

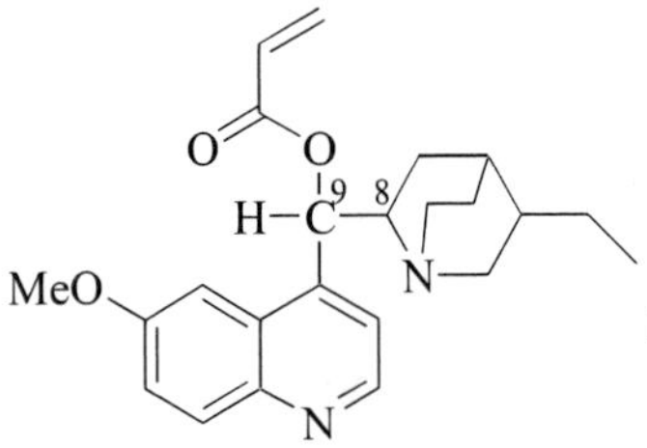

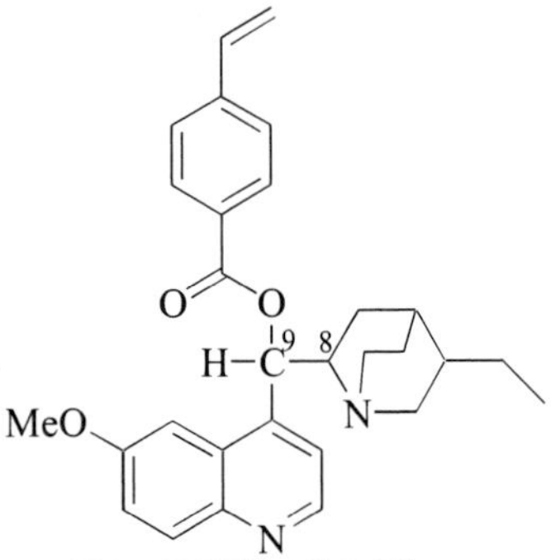

282a: DHQD: (8*R*,9*S*)
282b: DHQN: (8*S*,9*R*)

283a: DHQD: (8*R*,9*S*)
283b: DHQN: (8*S*,9*R*)

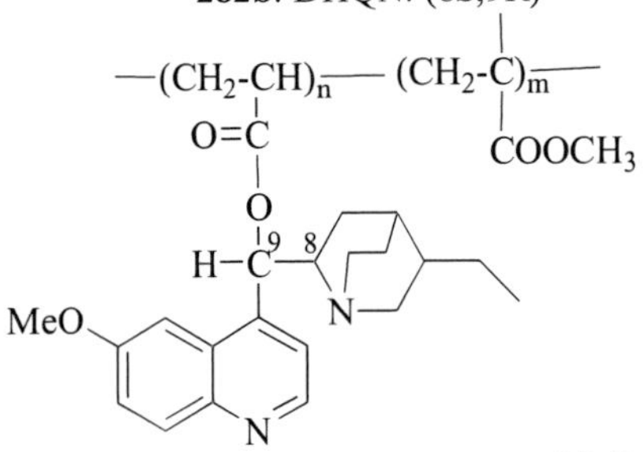

284: n=1, m=0, (8*R*,9*S*)

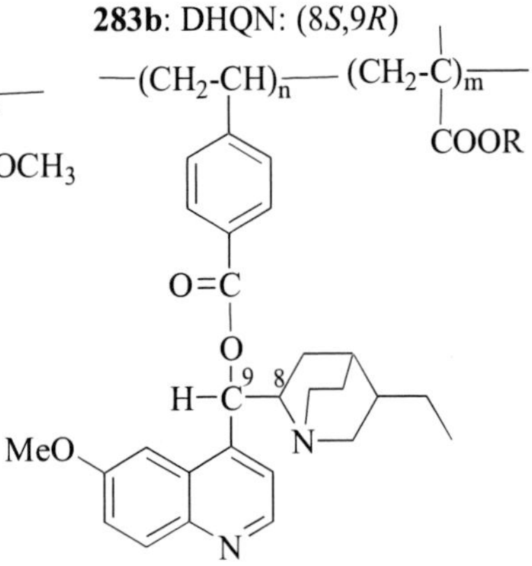

285a: n=1, m=0, (8*R*,9*S*)
285b: n=1, m=0, (8*S*,9*R*)
286a: R=CH$_3$, n=0.2, m=0.8, (8*R*,9*S*)
286b: R=CH$_3$, n=0.2, m=0.8, (8*S*,9*R*)
287a: R=CH$_2$CH$_2$OH, n=0.06, m=0.94, (8*R*,9*S*)

Scheme 114.

These two polymer-supported alkaloids were employed in the AD of *trans* metyl cinnamate and *trans* stilbene using OsO$_4$ as oxidant and K$_3$Fe(CN)$_6$-K$_2$CO$_3$ in *t*-BuOH/H$_2$O (1/1) at 10°C. Excellent enantioselectivities (>98%) were observed in all cases. The OsO$_4$/**289a** complex was easily filtered off from the reaction mixture and reused a second time in the dihydroxylation of *trans* stilbene. Although the enantiosectivity remained excellent (99%), the reaction rate was slower (65% yield after 32h *vs* 93% yield after 15h). In opposite, the polymer **289b** was difficult to filter because of its high swellability.

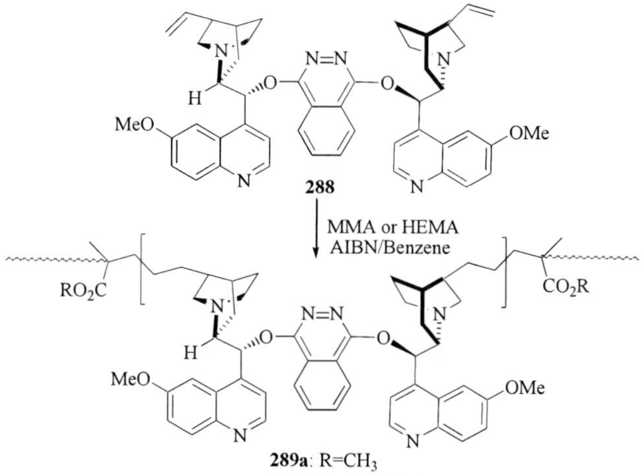

288

MMA or HEMA
AIBN/Benzene

RO₂C — CO₂R

289a: R=CH₃
289b: R=CH₂CH₂OH

Scheme 115.

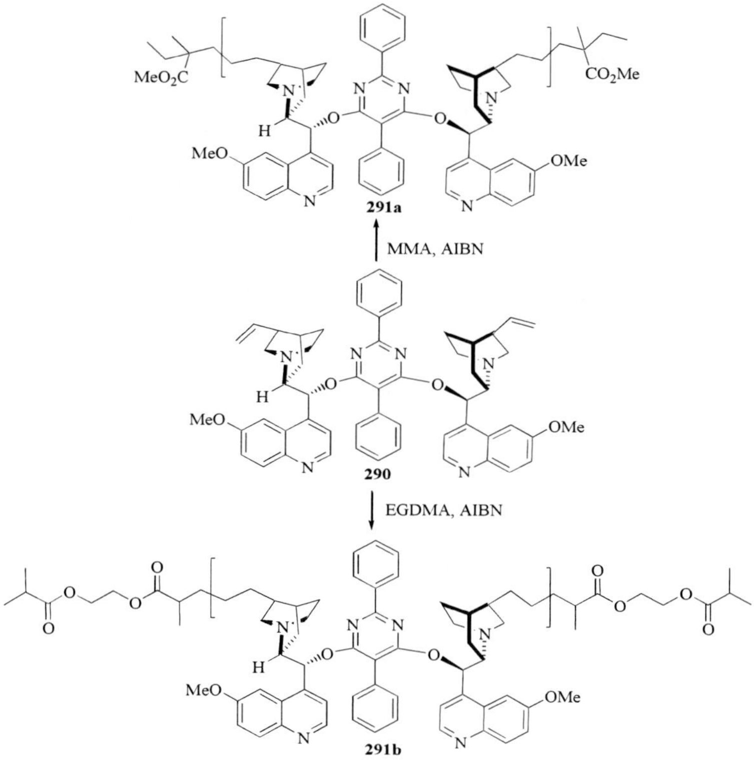

291a

MMA, AIBN

290

EGDMA, AIBN

291b

Scheme 116.

An analog of monomer **288** has been prepared replacing the phthalazine by a 2,5-diphenylpyrimidine and the corresponding monomer **290** was copolymerized with methyl methacrylate or ethylene glycol dimethacrylate to lead to the copolymers **291a** and **291b** respectively (Scheme 116) [177].

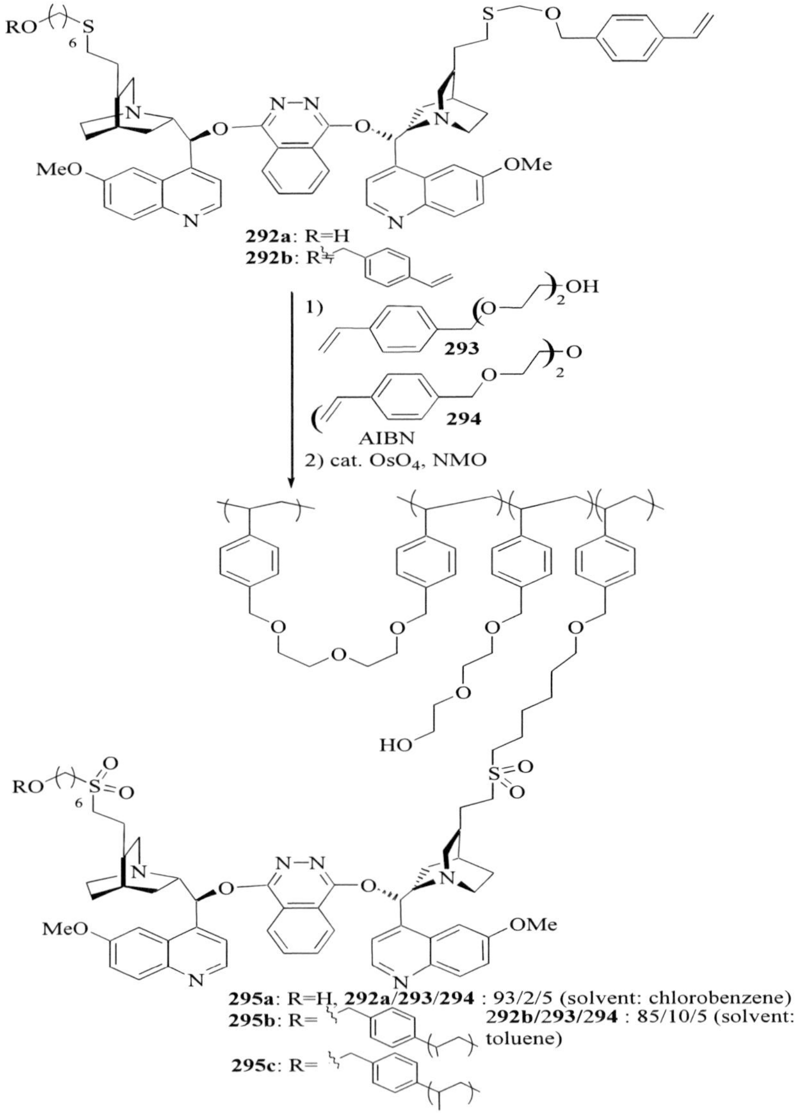

292a: R=H
292b: R=

1)

293

294

AIBN
2) cat. OsO$_4$, NMO

295a: R=H, **292a/293/294** : 93/2/5 (solvent: chlorobenzene)
295b: R=   **292b/293/294** : 85/10/5 (solvent: toluene)

295c: R=

Scheme 117.

Unfortunately, the enantioselectivities were around 15-45% lower than those observed with the phthalazine-derived supported dihydroquinidine **289**.

Another type of insoluble polymer-bound cinchona alkaloid was tested in the AD of olefins. Salvadori has prepared copolymers **294** by solution or suspension copolymerization of chiral monomers **292** with the two styrenic compound **293** and **294** with different monomer ratio (Scheme 117) [178].

Both yield and ee were very high and copolymer **295a** proved to be more efficient than **295b** and **295c**. The better swellability due to the lower crosslinking degree has certainly more influence on the catalytic properties and the enantioselectivity than the solvent or the technique polymerization. The supported alkaloid could be recovered and reused several times without significant loss of the enantioselectivity.

7.2. SOLUBLE SUPPORTED ALKALOID AS LIGAND

The immobilization of a ligand on a MeO-PEG leads to the corresponding soluble polymer and it is known that the solubility could have an influence on the outcome of the reaction.

For this reason, soluble polymer **296** was prepared and employed in the AD of olefins (Scheme 118) [179]. Similar yield and ee were observed for the reaction with *trans* stilbene than those obtained with the insoluble polymer-supported alkaloid **262** but with this soluble polymer the reaction time was very short (5h *vs* 48h). With other olefins used, the yield and ee were less satisfactory. The addition of diethylether allowed the precipitation of the supported alkaloid which could be reused without decrease of the catalytic activity and enantioselectivity.

The soluble polymer containing the highly active ((DHQD)$_2$-PHAL) **297** was also prepared and employed in the dihydroxylation of olefins (Scheme 118) [180]. The same level of enantioselectivitity (> 97% for all alkenes) than that obtained with the insoluble analog **289a** was observed.

Bolm has reported the first anthraquinone-derived supported dihydroquinidine for the dihydroxylation reaction [181]. Two different ways about attaching the anthraquinone ligand were studied: either *via* the anthraquinone part or *via* the double bond of the natural quinidine (Scheme 119). The two families of soluble supported bisdihydroquinidine **298** and **299** were prepared and tested in the AD of allyl iodide and indene using 0.4 mol% of K$_2$[OsO$_2$(OH)$_4$] as oxidant and K$_3$Fe(CN)$_6$-K$_2$CO$_3$ as cooxidant in *t*-BuOH/H$_2$O (1/1).

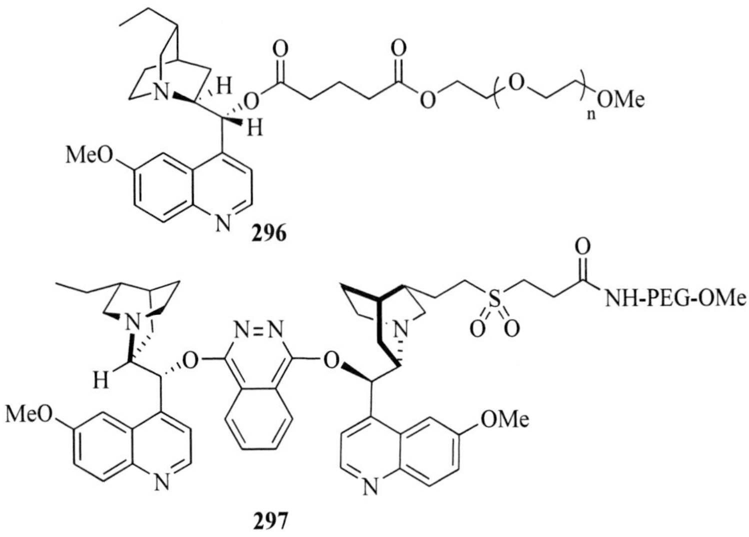

296

297

Scheme 118.

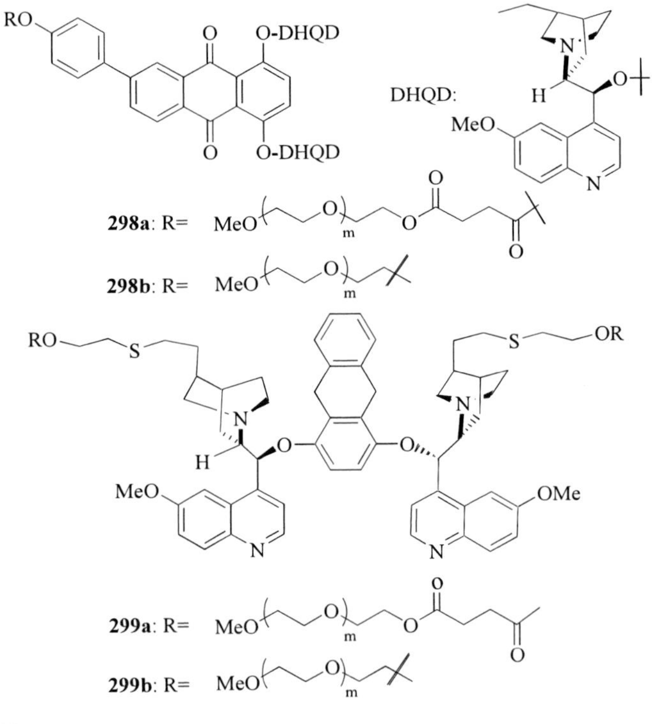

298a: R=

298b: R=

299a: R=

299b: R=

Scheme 119.

In the case of allyl iodide, the soluble supported ligands gave the same enantioselectivities than the non-polymeric ligand. Although the yields were better for the AD of indene, the ee were slightly lower for all the supported ligands employed. The anchoring or the binding site had absolutely no influence on the enantioselectivity. After addition of MTBE, the supported alkaloid **298a** and **299a** could be recovered and reused in another dihydroxylation reaction and no loss of performance was detected.

This type of supported anthraquinone-derived alkaloid has been also applied in a continuous flow Sharpless dihydroxylation using the continuously run chemzyme membrane reactor (CMR) in which the catalyst attached to the soluble polymeric ligand **300** is retained by a membrane (Scheme 120) [182]. The two cooxidant systems have been tested and although the NMO one gave lower ee, this system was more compatible with the CMR. The different solutions were delivered in the reactor and after a residence time of 85 minutes, fractions from the reaction mixture were collected. At the beginning, both conversion and ee were around 80% and after only six residence times the conversion dropped to 18% suggesting that a leaching of osmate occurred.

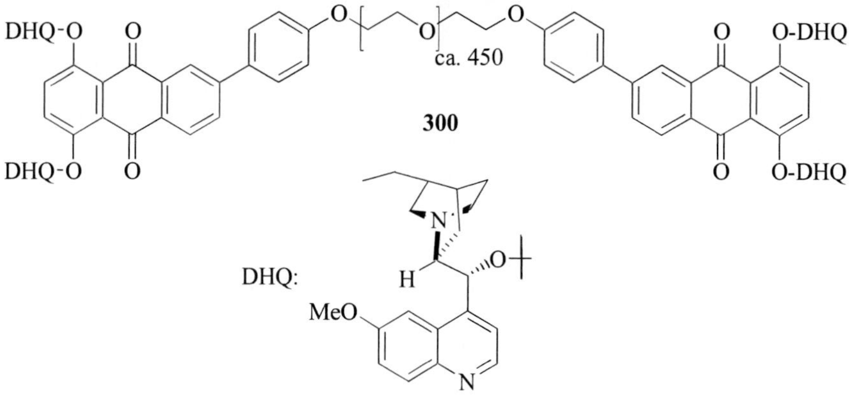

Scheme 120.

Excellent results were obtained for the asymmetric aminohydroxylation using the soluble polymer-bound bisdehydroquinine **301**, prepared in two steps from dehydroquinine (Scheme 121) [183]. When analogs of *trans* cinnamate **278** were used using $K_2[OsO_2(OH)_4]$ as oxidant, AcNHBr as nitrogen source in the presence of LiOH in *t*-BuOH/H_2O (1/1), regioselectivity (>20/1) and ee (> 95%) were excellent particularly with **301b** and **301c**. Results were less satisfactory when the acetone/water was used as solvent. After precipitation and filtration, the PEG-bound ligand **301c** could be reused in five consecutive runs with an average decrease of 0.8% per run for the enantioselectivity.

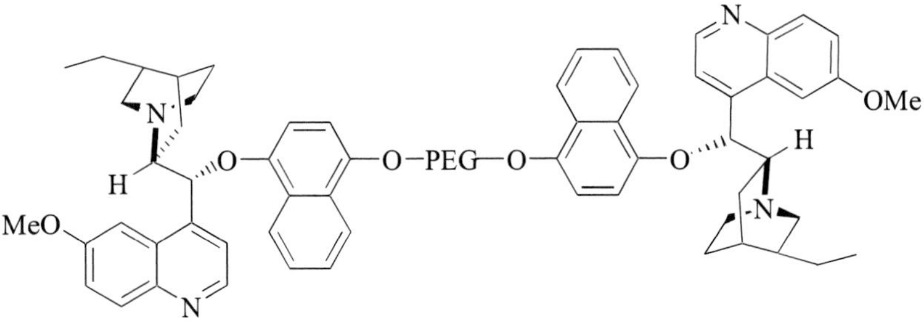

301a: HO-PEG-OH, MW *ca.* 4000
301b: HO-PEG-OH, MW *ca.* 6000
301c: HO-PEG-OH, MW ca. 8000

Scheme 121.

Anthraquinone-derived quinine has also been anchored onto PEG-OMe to produce soluble polymeric alkaloid **302** (Scheme 122) [184]. Good to excellent yields (80-94%) and ee (80-99%) were observed with different olefins using $K_2[OsO_2(OH)_4]$ as oxidant, $K_3Fe(CN)_6$-K_2CO_3 as cooxidant in *t*-BuOH/H_2O (1/1).

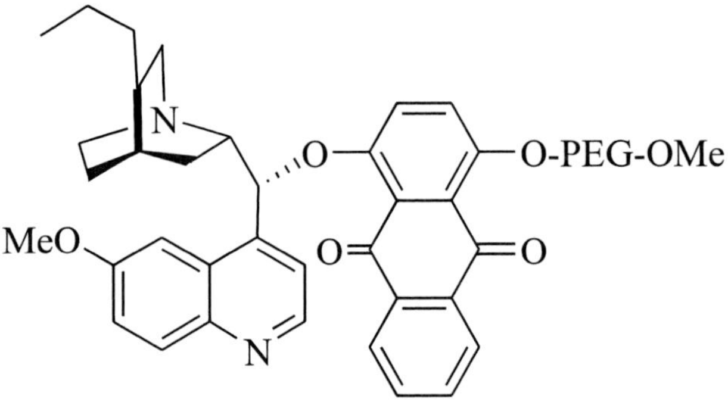

302

Scheme 122.

Chapter 8

EPOXIDATION

Epoxides constitute a class of versatile intermediates, as they can be easily transformed into a wide variety of functional groups involving regioselective ring opening reactions [185]. Thus, asymmetric epoxidation (AE) of olefins is a key reaction for the synthesis of enantiomerically enriched compounds.

Complexed with Mn, Cr or Co, *N,N*-ethylenebis(salicylidene aminato) derivatives known as salen compounds, first described in 1990 by Jacobsen, are widely used as catalysts. From them, manganese complexes have been reported as highly active and enantioselective catalysts in the epoxidation of unfunctionalized alkenes in homogeneous phase using a wide range of oxidants [186]. Number of structure presenting the same backbone have been synthesized and involved in these oxidation reactions (Scheme 123). Easy to handle and to obtain, rather cheap, uses of Jacobsen type catalysts represent one of the most relevant methods of building chiral epoxides. Immobilization of such a compound was extensively studied because it restrains the formation of μ-oxoMn(IV) dimers by isolation of the catalytic sites.

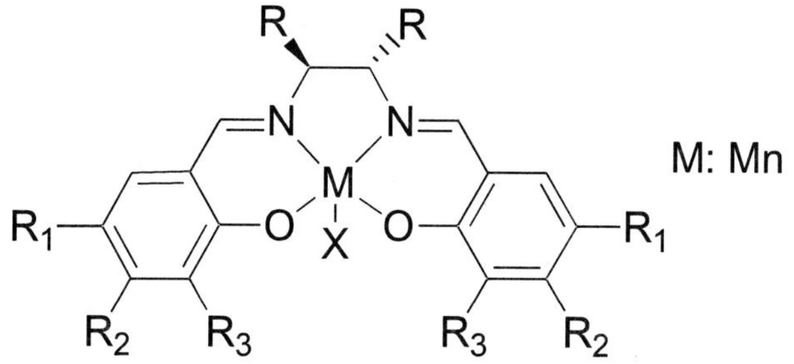

Scheme 123.

The immobilization of salen onto an organic polymer was achieved according to two strategies: a copolymerization of a salen monomer or by grafting onto a polymer followed by subsequent solid phase synthesis of the salen. In the latter case, only unsymmetrical salen are formed.

8.1. SUPPORTED SALEN AS LIGAND

The first example of a polymer-supported Jacobsen catalyst was reported by Dhal [187]. It involved the radical copolymerization of the corresponding distyryl monomer with EGDMA in a ratio of 10/90 to give the macroporous polymer **303** (Scheme 124). The epoxidation of unfunctionalized olefins led to moderate yields (55-72%) and low ee (up to 30%). Moreover, this catalytic system could be used at least five times without any significant loss of its catalytic activity.

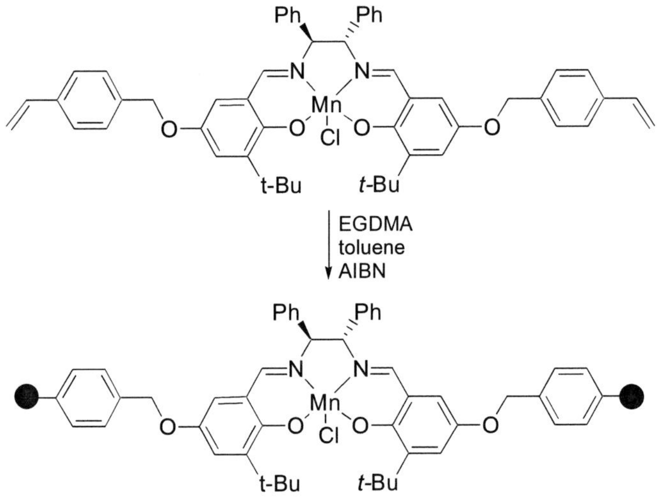

Scheme 124.

A similar approach permitted to prepare a macroporous polystyrene-based polymer **306** *via* a radical copolymerization of a divinylsalen **304** or of a "distyryl spaced" salen **305** with styrene and DVB (ratio 10/75/15) (Scheme 125) [188].

The oxidation was performed with *m*-CPBA/NMO instead of PhIO as oxidant because this latter was transformed into $PhIO_2$ its disproportionation product which was insoluble in the medium. The catalytic reaction was performed with 10 mol% of the catalyst at 0°C. After 30 minutes, modest to excellent conversions (67-99%) with low to moderate ee (10-62%) were observed. Moreover, the catalytic system could be recycled 5 times, the activity and the selectivity remained identical. Compared to homogeneous systems, enantioselectivity was still low. The epoxide of styrene, *cis*-β-methylstyrene and indene were formed in 16%, 62% and 60% ee respectively with conversions up to 99% in each case with polymer **305a** as ligand.

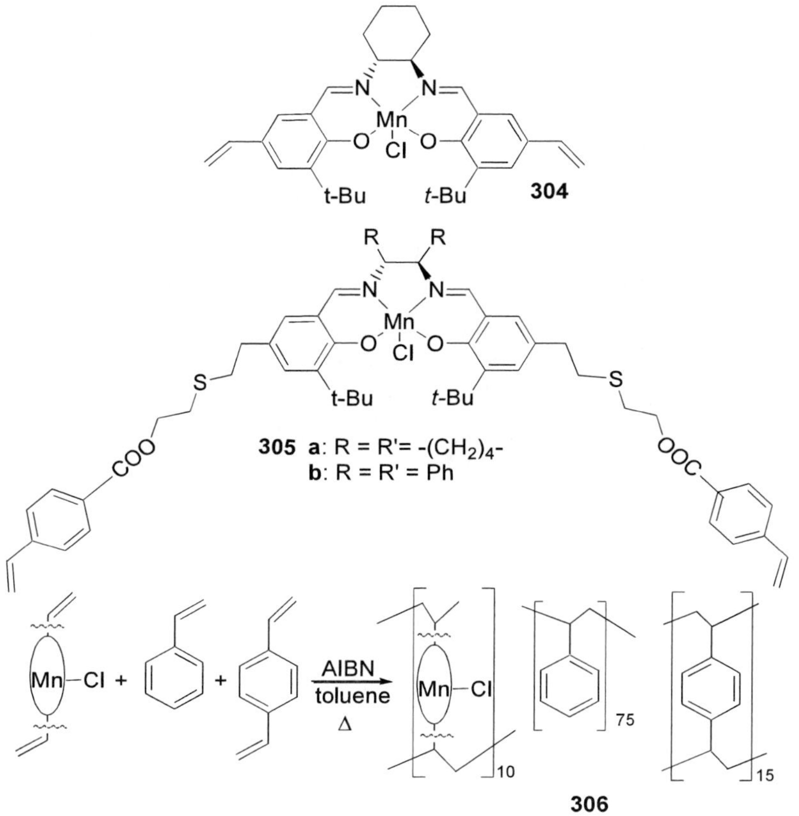

Scheme 125.

First it was considered that as the salen ligand was localized at the crosslink it may induced a steric hindrance and a conformational rigidity responsible of the low ee. Therefore, Laibinis [189-190] and Sherrington [191] synthesized non symmetrical immobilized salen attached specifically to the polymer in a pendent fashion. The polymer was a styrene base resin.

Laibinis used a three steps synthesis for the polystyrene-supported salen **307** (Scheme 126) [189-190]. Asymmetric epoxidation was carried out in biphasic conditions with NaOCl as the oxidant in the presence of **307$_{Mn}$** as catalyst. The enantiomeric excesses and the yields of AE with styrene (9% ee, 7% yield), *cis*-β-methylstyrene (79% ee, 2% yield) or dihydronaphtalene (42% ee, 46% yield) were low to modest.

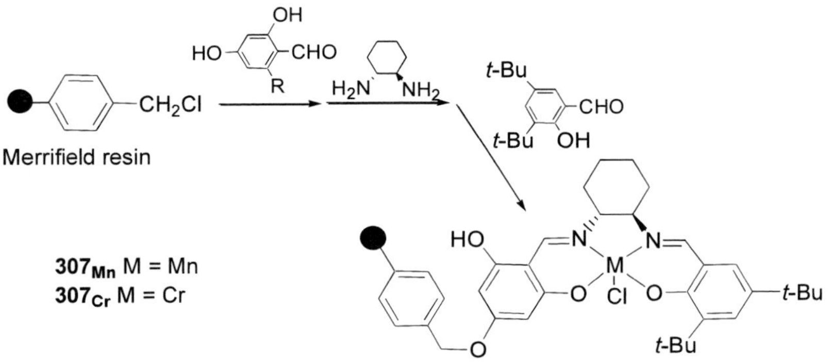

Scheme 126.

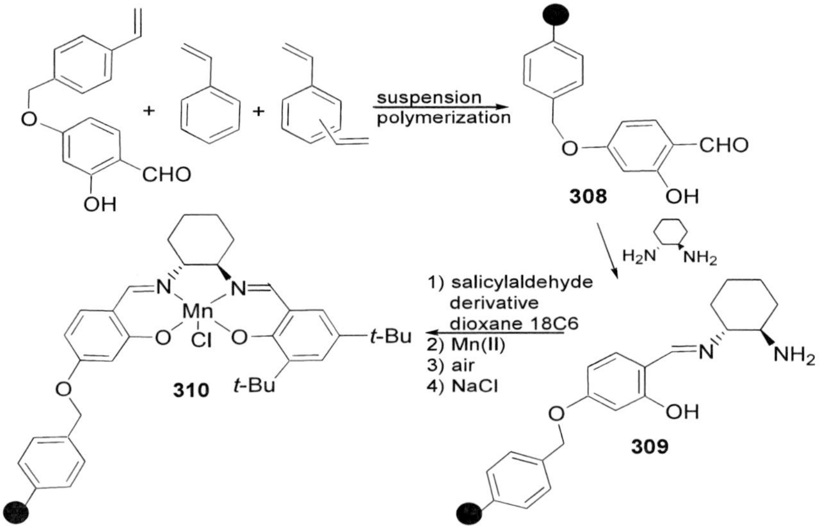

Scheme 127.

Sherrington [191] synthesized gel type and macroporous resins from 4-(4-vinylbenzyloxy)salicylaldehyde (Scheme 126). The gel type catalyst was rather inactive due to the formation of the oxo-bridged dimer of the catalytic center contrary to the macroporous one with an effective site isolation of the catalytic center. With the macroporous catalyst **310** (Scheme 127) although yields were good, ee were always low (up to 20% ee).

With an EGDMA polymer matrix (Scheme 128), polymer **311** afforded the epoxidation of 1-phenylcyclohex-1-ene with a high ee (91%) at 49% conversion. With the soluble Jacobsen catalyst, 92% ee and 72% conversion were observed [192-193].

The key factors for this result could be the low loading of Mn sites (0.08 mmol/g) and the high surface area.

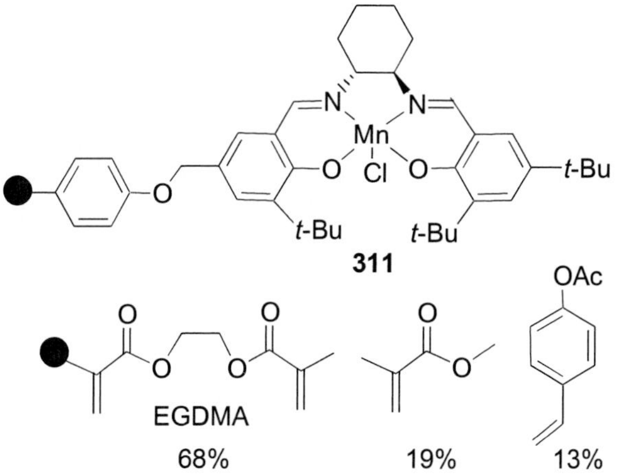

311

EGDMA
68%   19%   13%

Scheme 128.

From all these results a highly active and selective immobilized salen ligands should result in a molecular structure close to those of Jacobsen catalyst. Sherrington [193] thought also that to provide good catalytic performance of the polymer-supported catalyst the following design criteria could be met. The complex could be attached to a single flexible linkage to the polymer support in order to minimize the steric hindrance. Moreover, a low loading of the catalyst linked to the polymer was required to maximize site isolation of catalytic centres and hence to minimize the formation of the inactive oxo bridge dimer. The polarity of the resin should be such as to provide the optimum microenvironment for the catalyst and finally, the morphology of the resin must permit a high level of mass transfer to the active sites.

In 2000, taking into account the criteria edicted by Sherrington, Janda [194] described a convergent strategy to attach a salen ligand containing an appropriate linker to a low loading preformed polymer. A glutarate spacer (5 carbon atoms) was then used and the polymer **312** was prepared from styrene and polyTHF-derived crosslinker to form beads (Scheme 129). The AE in the presence of m-CPBA as oxidant was effective. Goods yields were obtain in the AE of styrene and cis-β-methylstyrene with 51% ee and 88% ee respectively. These results were similar to those achieved with the homogeneous Jacobsen catalyst. The supported catalyst could be reused 3 times without loss of activity and selectivity.

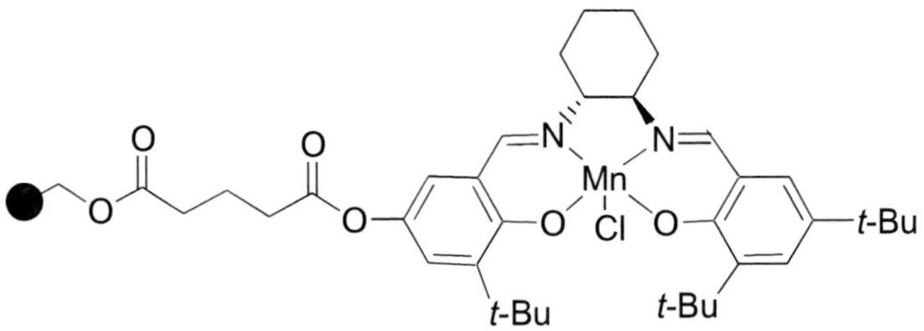

312

Scheme 129.

Another approach, keeping the C2 symmetry, consisted in the immobilization of the catalyst on a Tentagel amine resin via the pyrrolidine part of a pyrrolidine-salen instead of the aromatic rings (Scheme 130) [195]. In the presence of NaOCl or m-CPBA, with 4 mol% of catalyst **313** all underwent AE of 2,2-dimethylchromene, 6-cyano-2,2-dimethylchromene and 1-phenylcyclohex-1-ene in high yields, more than 70% and with enantio-selectivities of 82%, 86% and 68% respectively. In these conditions, decomposition of this catalyst was observed.

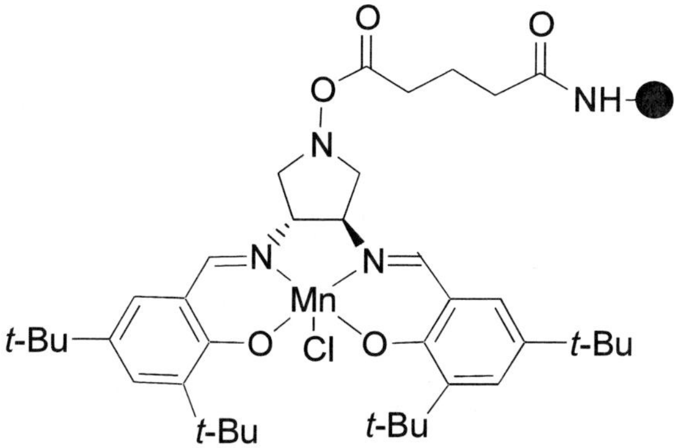

313

Scheme 130.

To conserve the C2 symmetry, Zhen [196-197] formed poly-salen complexes by condensation of a slight excess of chiral diamino cyclohexane with subsequent insertion of Mn ion (Scheme 131). These catalysts were soluble in CH_2Cl_2 but almost insoluble in ether. The oxidation reaction was performed in dichloromethane. To avoid any degradation of the polymer NaOCl was used in the presence of 4-phenylpyridine-*N*-oxide (4-PPNO) or *m*-CPBA in the presence of NMO as cooxidants. With *m*-CPBA/NMO enantioselectivities and activities are higher than those obtained with NaOCl/4-PPNO. With salen **314a** or **315** (Scheme 131), the catalytic system was similar to Jacobsen's one; at 0°C, up to 45% ee were observed for the AE of styrene. The lower temperature, the higher catalytic efficiency. For styrene epoxide formation, at room temperature and at -78°C the yields were 61% and 91% respectively with no significant improvement of the enantioselectivity of about 40%. Recovery and reuse of the catalytic system by precipitation was possible with only slight loss in enantioselectivity. The behaviour of catalyst **314b** (Scheme 131) was a little different [197]. At -78°C, it presented much lower activity but at -22°C, 83% yield and 51% ee were observed which was represented the best ee for the epoxidation of styrene. This catalytic system unstable in the reaction conditions was not reused.

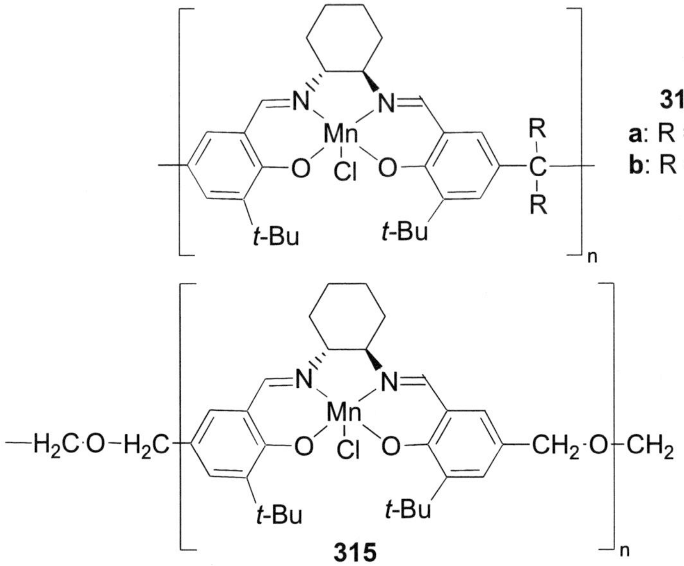

Scheme 131.

Kureshy [198] had also carried AE with a poly-salen. The poly-salen **316** derived from chiral diphenylethylenediamine (2 mol%) (Scheme 132) and NaOCl in the presence of cooxidant (NMO, 4-PPNO, PyNO, DMSO) forming the oxidative reagent were used for the epoxidation of styrene, indene and 6-cyano-2,2-dimethylchromene. In all cases, the best results in term of activity (yields up to 99%) and enantioselectivity were obtained with PyNO as cooxidant. The epoxide of styrene, indene and 6-cyano-2,2-dimethylchromene were formed in 32%, 78% and 100% ee respectively. Comparatively, poly-salen **314a** (n = 12, Scheme 131) led to 55% ee for epoxystyrene but for epoxyindene and epoxy 6-cyano-2,2-dimethylchromene 69% and 100% ee were observed. The turn over frequency (TOF) was higher with poly-salen **314a** (Scheme 131). The activity of the recycled catalyst gradually decreased upon successive use possibly due to minor degradation under epoxidation conditions and/or weight loss during recovery process of the chiral catalyst, but the ee remained constant.

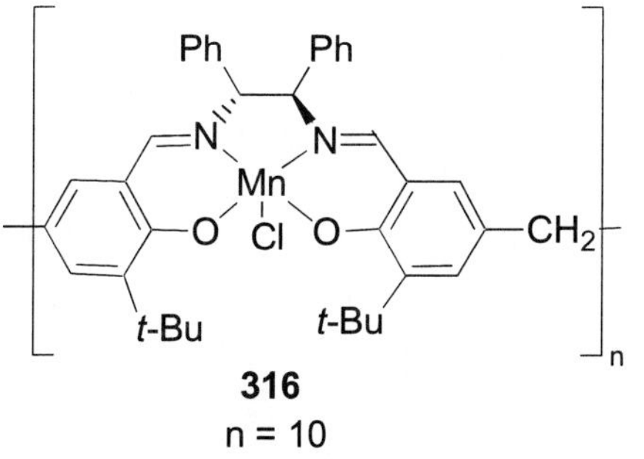

316
n = 10

Scheme 132.

Smith [199] involved immobilized Katsuki-type (salen)Mn complex **318** (R = H, Scheme 133) for the AE of 1,2-dihydronaphtalene with NaOCl/4-PPNO. The Katsuki-type (salen) **317** presenting an hydroxyalkyl group (R = $CH_2CH_2CH_2OH$ Scheme 133) on the 6-position was grafted onto a polystyrene resin *via* an ester link by means of polystyrene carboxyl chloride (1% crosslinked, 200-400 mesh, 1.17 mg/mol of chloride) (salen **319**, Scheme 133). The counter ion seemed to have no influence on the catalytic efficiency and selectivity. For all of them, yields were about 40% with excellent ee (upper than 90%). Each catalytic system was reused after filtration, ee and yields remained quite unchanged. To increase the yield upon the recycle (up to 70%), concentration of the reaction mixture was required.

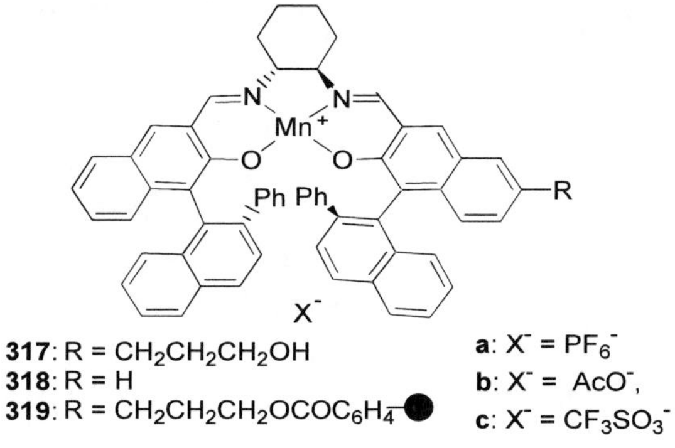

317: R = $CH_2CH_2CH_2OH$ **a**: $X^- = PF_6^-$
318: R = H   **b**: $X^- = AcO^-$,
319: R = $CH_2CH_2CH_2OCOC_6H_4$●   **c**: $X^- = CF_3SO_3^-$

Scheme 133.

The molecular imprinting polymer (MIP) was tested by Gohdes [200]. The manganese salen precatalyst **320** (Scheme 134) having two polymerizable acryloyl groups could be covalently embedded in the crosslinked polymer network. In order to mimic the bound styrene substrate, phenylacetate ligand was fixed on the Mn atom which after its removal will leave behind the substrate-binding cavity. The MIP synthesis was performed with the precatalyst **320** in the presence of EGDMA as the crosslinking agent, styrene as a non linking polymer and dichloroethane as a porogen. Exploration of porosity and crosslinking effects on activity on epoxidation of the styrene in the presence of PhIO or *m*-CPBA was carried out. In the optimum conditions 40-50% porogen and 70% crosslinking and *m*-CPBA as oxidant, the ee was up to 14%. This imprinted polymer could be reuse at least 3 times without loss of ee.

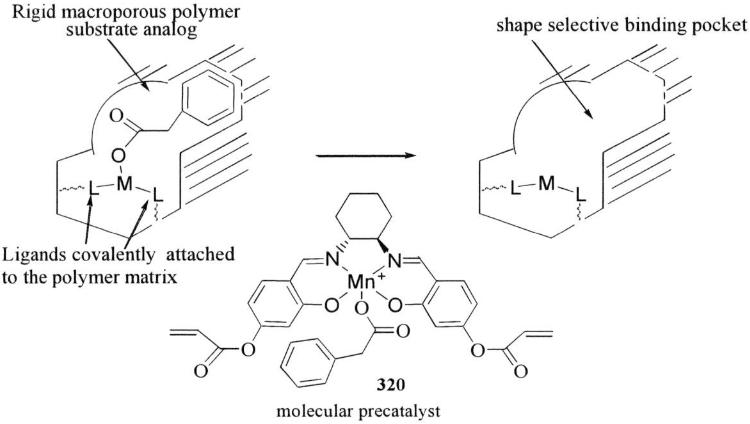

Scheme 134.

Another way to immobilize the salen catalyst was to use glucose as a linker and to graft it on CHO modified Wang resin (Scheme 135) [201]. The activity of the catalyst **322** in the AE of styrenes has been investigated with four oxidants: H_2O_2, NaOCl, (*n*-Bu$_4$N)HSO$_5$ and *m*-CPBA. The best results in terms of activity (99%) have been achieved with *m*-CPBA/NMO; the respective ee were 30% and 80% for styrene and *cis* β-methylstyrene. As the reaction was accompanied by extensive lack of metal from the solid, no recycle has been considered.

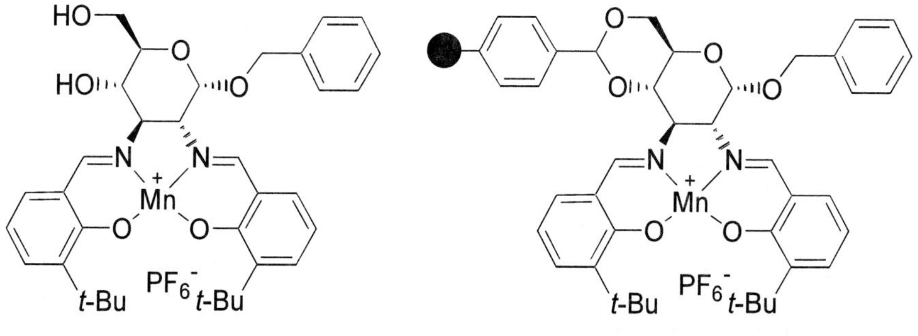

homogene catalyst
321

supported catalyst
322

Scheme 135.

A new approach consisting in the formation of rigid and amorphe salen-polymer (**323**, Scheme 136) was reported by Gothelf [202]. It involved a condensation of a rigid trisalicylaldehyde with substituted ethylene diamines (Scheme 136).

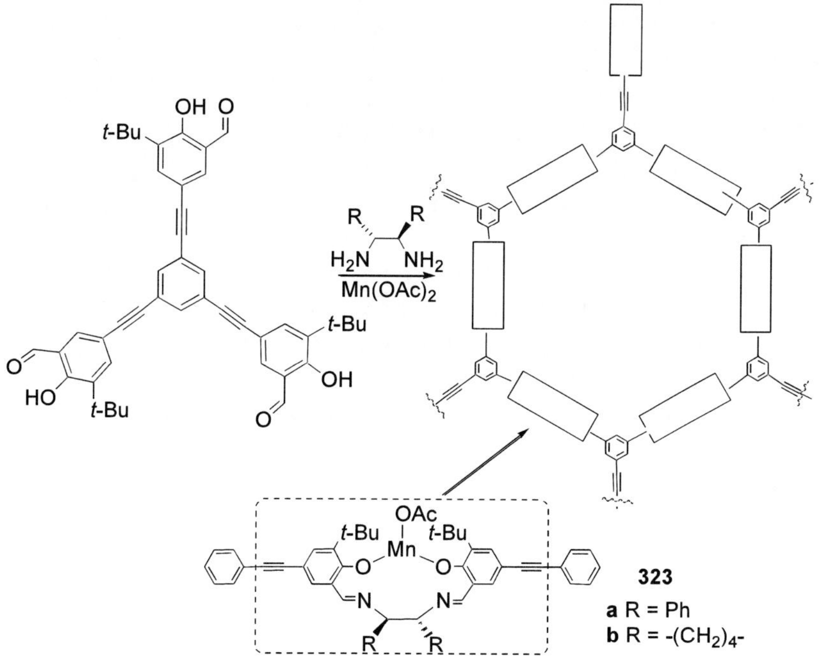

323
a R = Ph
b R = -(CH$_2$)$_4$-

Scheme 136.

The AE of *cis* β-methylstyrene at 0°C, with *m*-CPBA in the presence of NMO and polymer **323a** led to 78% conversion with a cis/trans epoxide ratio of 17.0, with a catalyst loading of 15 mol%. Up to 67% ee for the *cis* stereoisomer and 23% ee for the *trans* diastereoisomer are obtained. After isolation from the reaction mixture, the catalyst was reused at least 6 times without significant drop in reactivity and selectivity.

Weck [203] developed a monomer salen complex linked to a norbornene *via* a stable phenylene-acetylene linker and its subsequent polymerization by means of the controlled ROMP method using 3rd generation Grubb's catalyst (Scheme 137). This polymerization methodology led to fully functionalized immobilized metal-salen catalyst. By this way, the supported catalyst showed catalytic activities and stereoselectivities similar to the non-supported Jacobsen catalyst. Moreover, activities and selectivities seemed to depend on the density of the catalytic moieties: homopolymer **324** were less selective than their copolymer analogs **325**. For example, AE of 1,2-dihydronaphtalene led in both cases to total conversion and 76% ee for the homopolymer **324** *vs* 81% ee for copolymer **325a**. Recycle was possible and after 3 recyles a drastic decrease in ee was observed. AE of dihydronaphtalene led to 81% ee for the first cycle *vs* 6% ee for the third one.

8.2. PORPHYRINS

Polymer RuCO-porphyrins **102** and **103** (Scheme 48) were used for the cyclopropanation reaction of styrenes but they were also tested in the epoxidation reaction. Contrary to cyclopropanation for which moderate enantioselectivities and low activities were observed, these Ru porphyrin complexes gave good ee (up to 76%) and activity (up to 89%) for AE of unfunctionalized olefines [204]. Recently, Simmoneaux showed that iron catalyst **326** derived from electropolymerized tetraspirofluorenyl porphyrin (Scheme 138) led to moderate yields without chiral induction [205].

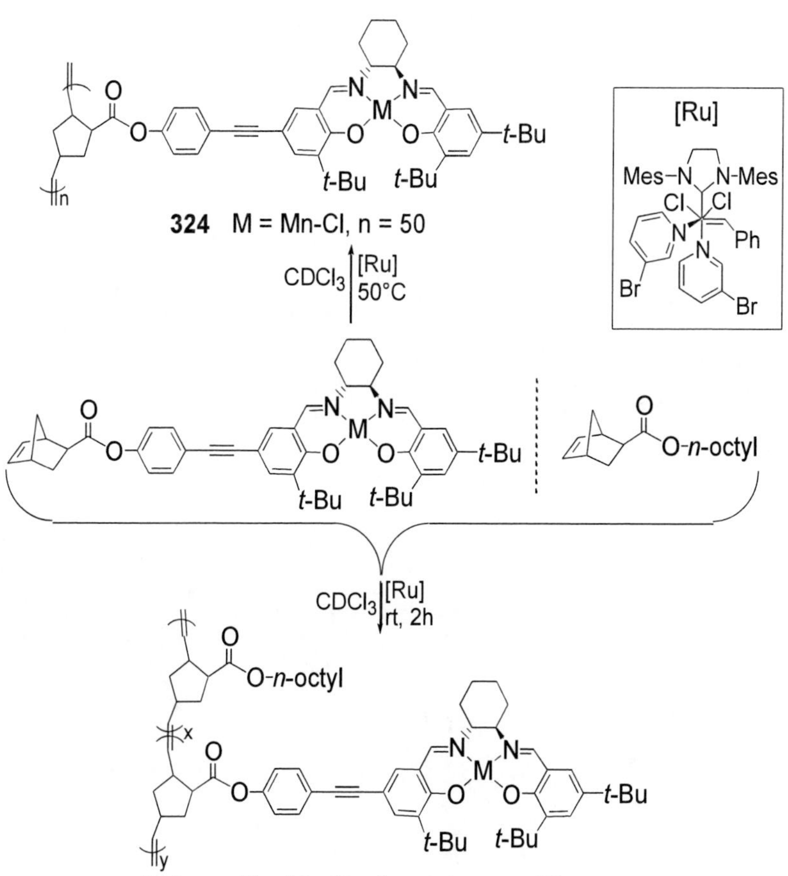

324 M = Mn-Cl, n = 50

[Ru]

CDCl$_3$ | [Ru] 50°C

O-*n*-octyl

CDCl$_3$ | [Ru] rt, 2h

O-*n*-octyl

325$_{Mn}$ **a** M = Mn-Cl, x/y = 1.1, x+y = 50
b M = Mn-Cl, x/y = 3.1, x+y = 100
c M = Mn-Cl, x/y = 9.1, x+y = 100
325$_{Co}$ **a** M = Co, x/y = 1.1, x+y = 50
b M = Co, x/y = 3.1, x+y = 100
c M = Co, x/y = 9.1, x+y = 100

Scheme 137.

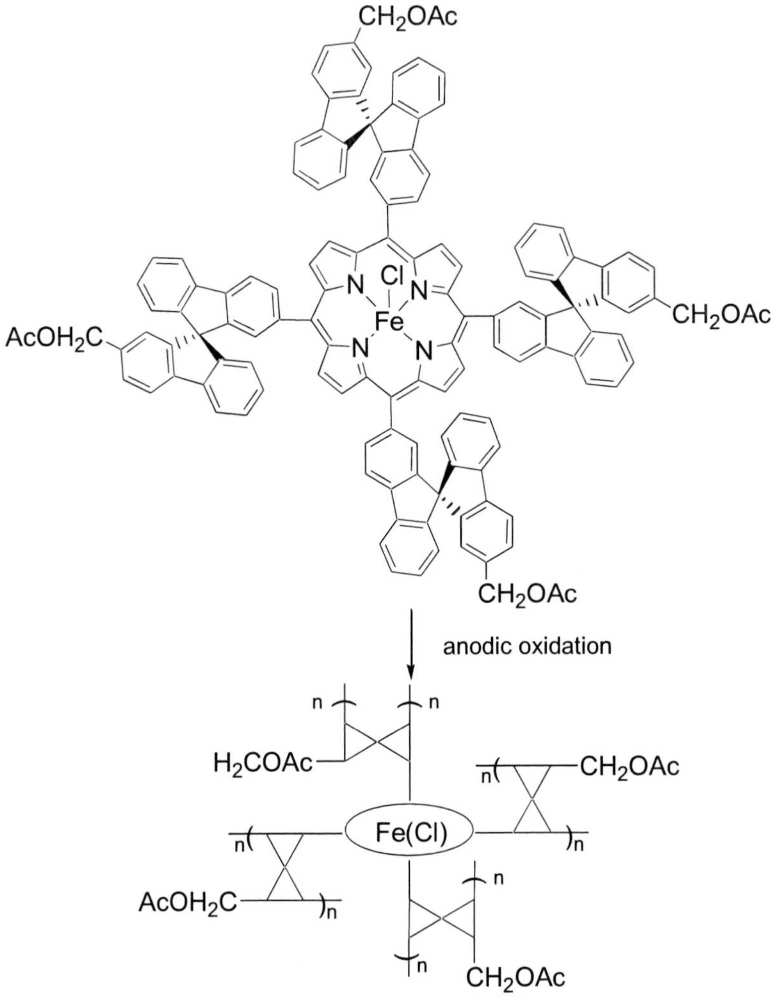

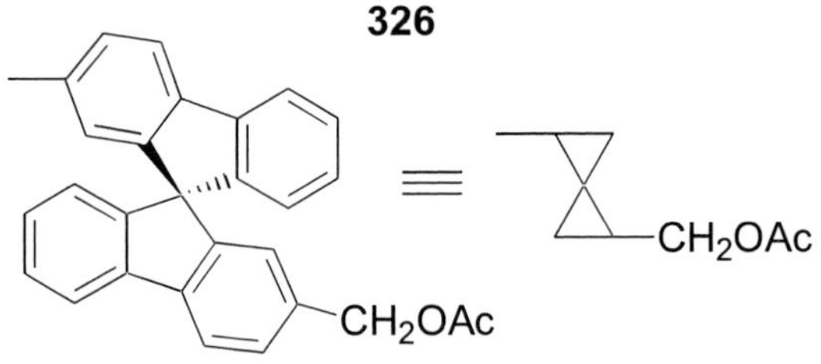

Scheme 138.

8.3. SOLUBLE POLYMERS

In 2006, Yu [206] prepared a polymeric chiral salen Mn(III) complex **327** containing chiral diamine spacer, acting as a solvent regulated phase transfer catalyst (Scheme 139). This catalyst presented a Mn ion content of 1.49 mmol/g and a molecular weight of 9100. Evaluation of its catalytic behaviour in AE of styrene was performed in the presence of the oxidative system *m*-CPBA/PyNO, *m*-CPBA/4-PPNO, *m*-CPBA/4-PPyNO or *m*-CPBA/NMO. Enantiomeric excesses up to 35% were found similar to those observed in the same reaction conditions with Jacobsen catalyst. This was probably due to the solubility of the polymeric catalytic system in the reaction mixture acting thus as an homogeneous catalyst by minimizing the diffusion of the reactants to the active sites of the catalysts. The nature and the amount of the *N*-oxide cocatalyst showed some impact on both yield and ee. Without this additive, the yield was rather low and no ee was detected; and its presence led to an increase of both yield and ee. The optimum yield and ee were obtained when 2 equivalents of each additive were used. But NMO was the best one leading to 98% yield and 43% ee. Recycle was possible but after the third one, poor yields and ee were observed.

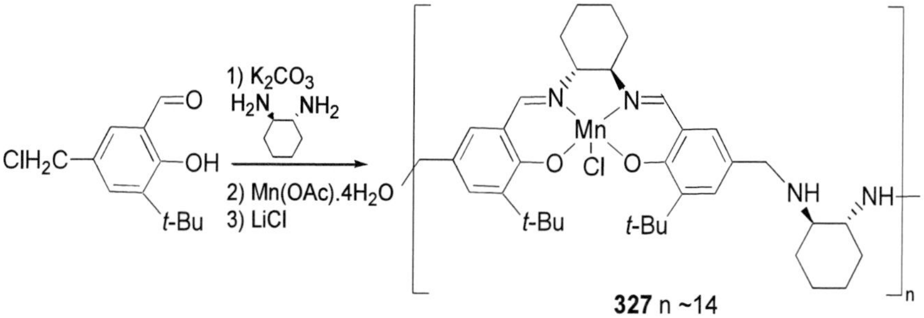

Scheme 139.

Recently, Liese [207-208] used a hyperbranched polyglycerol support as high loading Mn-salen complex (chemzyme) as a catalyst fo AE 6-cyano-2,2-dimethylchromene for a continuous application in a membrane reactor. This polymer supported catalyst was performed using a polyglycerol which molecular weight is 8000 g/mol and OH group loading 13.5 mmol/g. The polymer purification was carried out by ultrafiltration leading to 65% of the desired polymer **328** (Scheme 140). The following metal insertion led to polymer-supported catalyst **329**.

The oxidative reagent was formed by *m*-CPBA and NMO. Upon recycling it was observed an enhancement of the stability of the catalyst, the total turnover number increasing from 23.5 for a single batch to 80 in four repetitive batches. Moreover, some metal leaching occurred because the maximum conversion decreased from 98 to 75%, but enantioselectivities were less affected; they decreased from 95 to 88% ee from the first to the fourth batch. Nevertheless, in a continuously operated chemical membrane reactor, the TTN reached up to 240 after 20 residence times with conversions up to 70% and ee up to 92%.

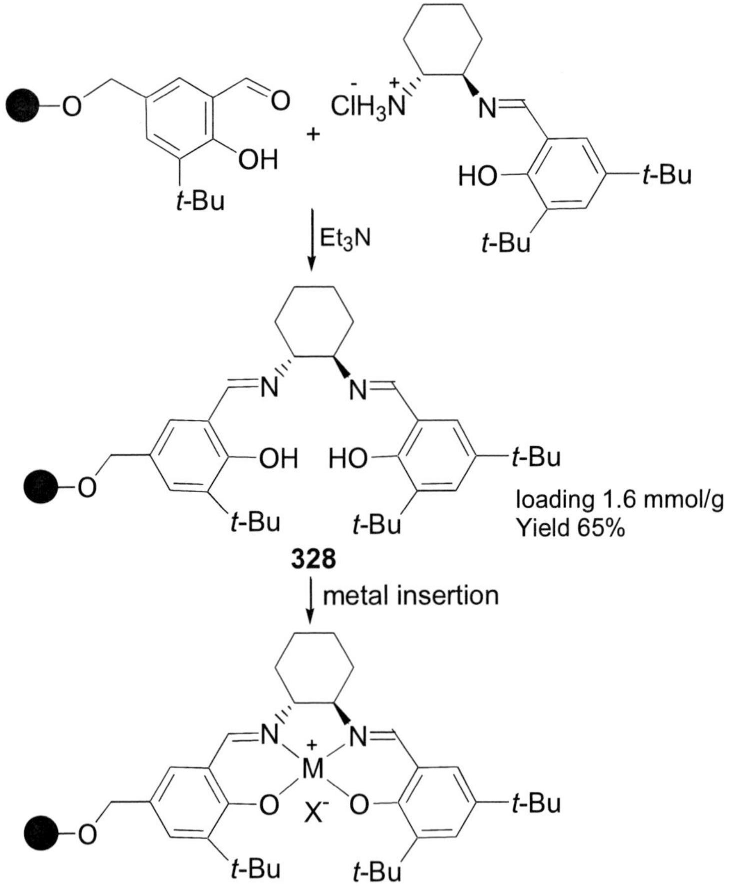

329$_{Co}$ M = Co, X = OAc, conversion = 54%, Co loading: 0.76 mmol/g
329$_{Mn}$ M = Mn, X = Cl, conversion = 75%, Mn loading 1 mmol/g

Scheme 140.

Chapter 9

KINETIC RESOLUTION
OF TERMINAL EPOXIDES

Close to Mn-salen complexed which were effective in the epoxidation of unfunctionalized alkenes Co or Cr-salen complexes have shown their ability in asymmetric ring opening such as hydrolytic kinetic resolution (HKR) of racemic epoxides. This reaction, performed in homogeneous conditions was able to furnish chiral epoxides and diols with high enantioselectivities. Numbers of inorganic supports have been used however some organic supports have also been studied. Other kinetic resolution could be performed such as ring opening epoxide with other nucleophiles and dynamic resolution.

Contrary to the catalytic systems employed for the epoxidation, mechanistic studies of asymmetric ring opening have shown that cooperative interactions between catalyst units are needed [209]. Thus, high local concentration of catalyst was necessary and high-loading support should increase the catalytic reactivity.

In 1999, Jacobsen [210] prepared a polystyrene-supported Co-salen complex **332** by grafting a monophenol derivative **330** of an highly efficient chiral salen onto hydroxymethyl polystyrene beads (90 μm) derivatized as their *p*-nitrophenyl carbonate **331** (Scheme 140). The same polystyrene complex **332** could be synthesized by resin capture of the salen **330**. The preparation of salen **330** was carried out using an excess of di-*tert*-butyl salicylaldehyde relative to 2,5-dihydroxy-3-*tert*-butylbenzyldehyde and enantiopure cyclohexane diamine (ratio: 3/1/2) yielding a 9/6/1 ratio of di *tert*-butyl, monophenolic (**330**) and diphenolic ligands. Then, addition of the carbonate derived polystyrene resin to the mixture allowed the selective capture of the two phenolic ligands. After washing, the polymeric mixture is formed by the incorporation of the diphenolic ligand in the resin bound ligand close to ligand **332**. Complexed with Co, this small amount of the incorporated diphenolic ligand in the resin catalyst did not affect the rate and the enantioselectivity; compared to material prepared with pure ligand **332**, similar results were observed.

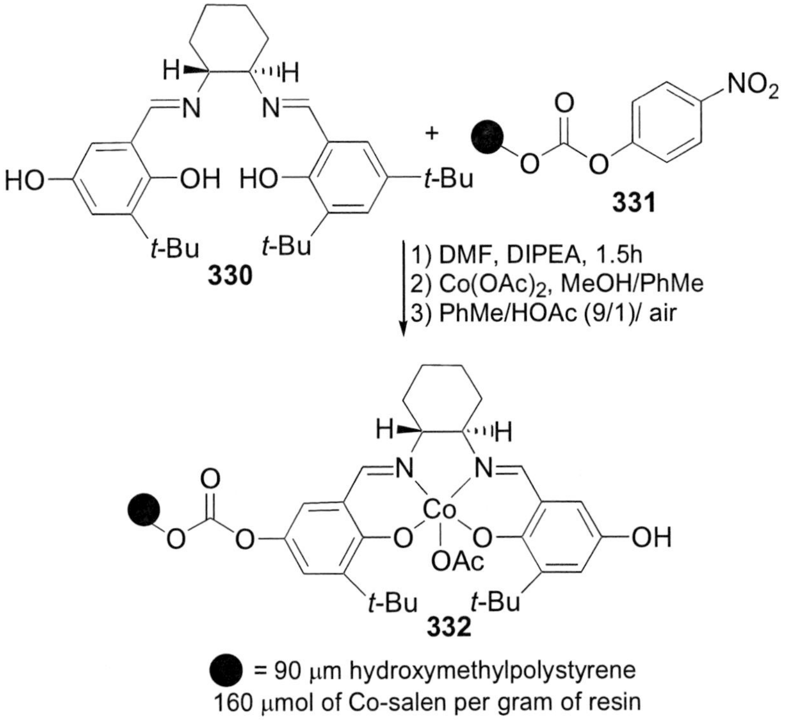

Scheme 141.

The HKR of epichlorhydrine (X = Cl, Scheme 142) and 4-hydroxy-1-butene oxide (X = OH, Scheme 142), the dynamic kinetic resolution of epibromhydrin and the enantioselective ring opening of epoxides by phenol were examined. In the first experiment, combination of the crude organic soluble products of the five recycle reactions and concentration led to the (*S*)-epichlorhydrine in 41% overall yield and > 99% ee and the (*R*)-chlorodiol in 93% ee. In the second one, the sum of five experiments provided (*S*)-triol in 36% overall yield and 94.4% ee while the enantioselectivity of the epoxide was only 59% ee.

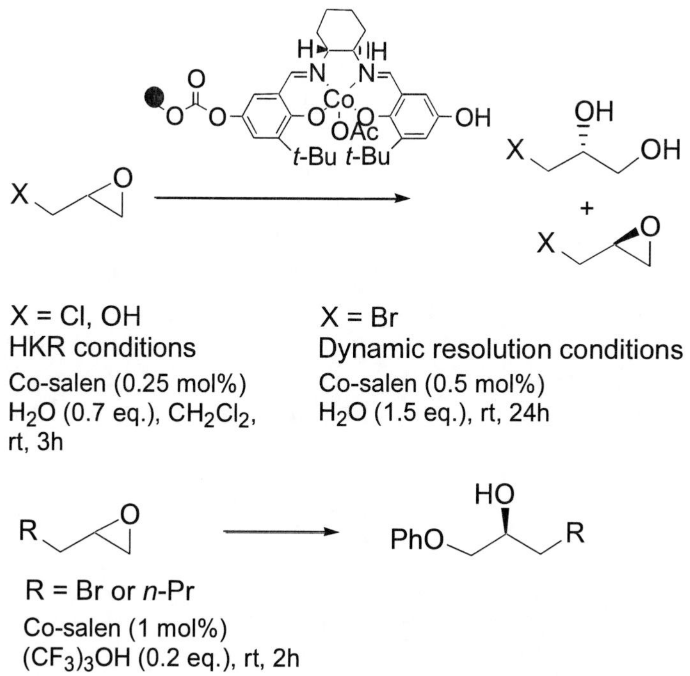

Scheme 142.

The dynamic resolution of epibromohydrine provided only the bromodiol in 94% overall yield and 96% ee (sum of the five experiments). Excellent results were also obtained for the ring opening epoxides with phenol ee could reach more than 99% and yield 98%.

The kinetic resolution of other epoxides with phenol was also studied in enantioselective parallel synthesis with the same catalyst, providing efficient access to important precursors of pharmacologically active compounds [211]. For all the studied substrates, the catalytic system was proved to be very efficient. β–aryloxyalcohols were formed in high yield (90-100%), purity ((93-100%) and ee (81->99%).

Zheng [212] prepared crosslinked polymeric Co(III)-salen complexes in order to induce cooperative effect. Trialdehyde **333** and dialdehyde **334** are formed *via* condensation of 2,5-dihydroxy-3-*tert*-butylbenzyldehyde with diacid and triacid derived from hydroquinone and chloroglycinol. Different proportions of tri and di aldehydes (tri/di) were subsequently condensed with enantiopure cyclohexane diamine to afford polymeric ligands **335** (Scheme 143). An average molecular weight between 4000 and 10000 were observed for these polymers. For the HKR of epichlorhydrine, styrene oxide, and phenylglycidyl ester, in a wide range of tri/di proportion, similar behaviour of the catalytic system was observed in the same reaction conditions. For example, with epichlorhydrine conversion based on racemic mixture varied from 50-52%, epoxide ee from 98-99% and diol ee from 91-97% using 0.02 mol% of the catalyst, at 10°C. Most of the crosslinked polymeric catalyst presented better activities than those of the oligomer **336**, showing a positive effect on the cooperation on catalytic centers with the crosslinked polymer. But with the complete crosslinked polymeric catalyst **335** activities and enantioselectivities are slightly lower, behaviour probably due to the poor solubility of the catalyst. Moreover, attempts to recycle the catalytic system were unsuccessful. This fact was ascribed to the sensitive ester linkage to the reaction medium.

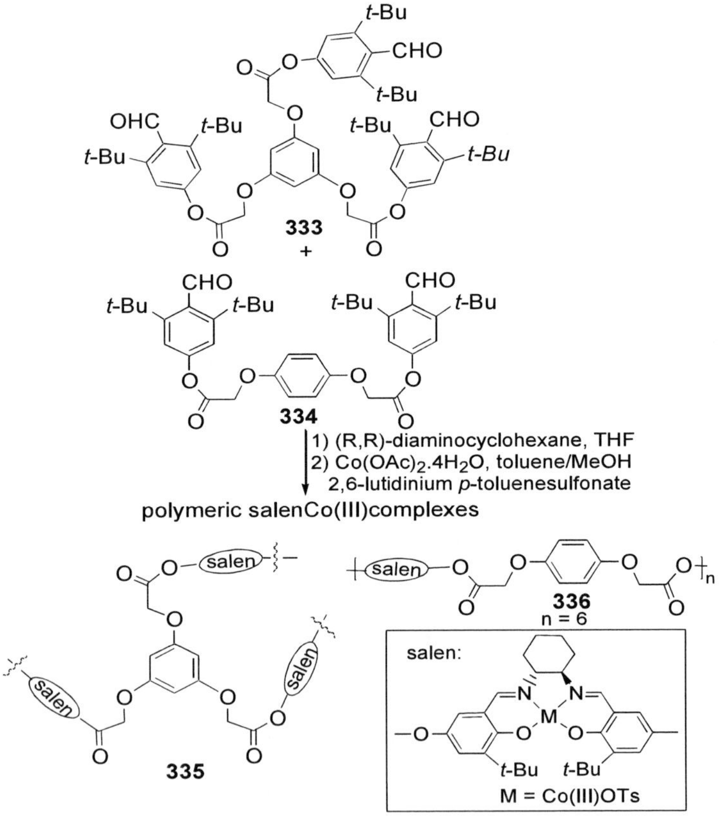

Scheme 143.

Kim [213] studied the effect of the counter ion on the HKR of epichlorhydrine, 1,2-epoxybutane, 1,2-epoxyhexane and epoxystyrene. The salen polymer catalyst were synthesized by copolymerization of salen bearing chloromethyl groups with sodium phenoxide derivatives of hydroquinone, 1,3,5-trihydroxybenzene or 1,1,1-tris(p-hydroxyphenyl)ethane in the presence of N-methylpyrrolidine and NaH in THF (Schemes 144 and 145). Co(II) type polymeric chiral salen ligands were formed by reaction of the corresponding salen ligand with hydrous Co(II) acetate. To obtained Co(III) polymeric chiral salen catalysts, the Co(II) polymer salen was treated with ferrocenium hexafluorophosphate or ferrocenium tetrafluoroborate noted respectively Co(III).(PF$_6$), Co(III).(BF$_4$). The HKR of epichlorhydrine, 1,2-epoxybutane, 1,2-epoxyhexane and epoxystyrene was performed. With all these substrates, excellent enantioselectivities (up to 99.8%) were observed with **337-**, **338-**, **339-**, **340-**Co(III).(PF$_6$), -Co(III).(BF$_4$) catalysts. Both type of catalysts exhibited almost the same ee as monomeric salen catalyst. These catalytic systems were stables; they could be recovered and reused at least seven times without further treatment without loss of activity and selectivity and without regeneration of the catalyst. However, it is noteworthy that **340-**Co(III)OAc catalyst, the regeneration with AcOH/air is necessary to avoid an ee decreased to 20% in the second hydrolysis. Moreover with **340-**Co(III)OAc catalyst a racemization during the reaction and the product isolation by distillation could occurred. The ee decreased slowly *via* racemization which is not the case with **340-**Co(III).(PF$_6$).

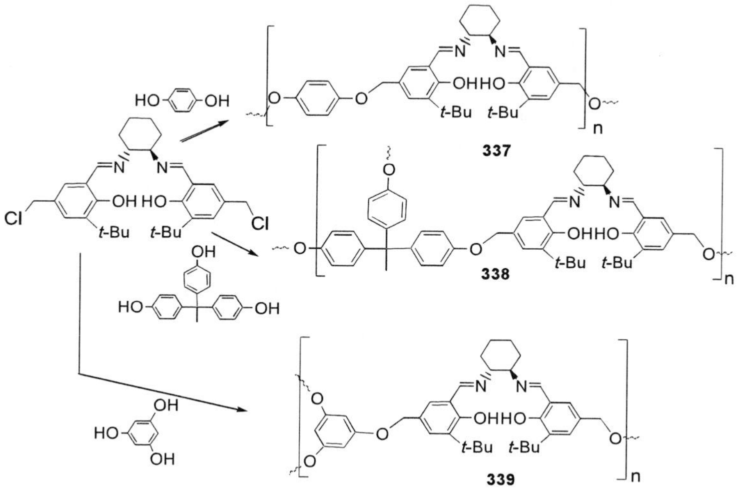

Scheme 144.

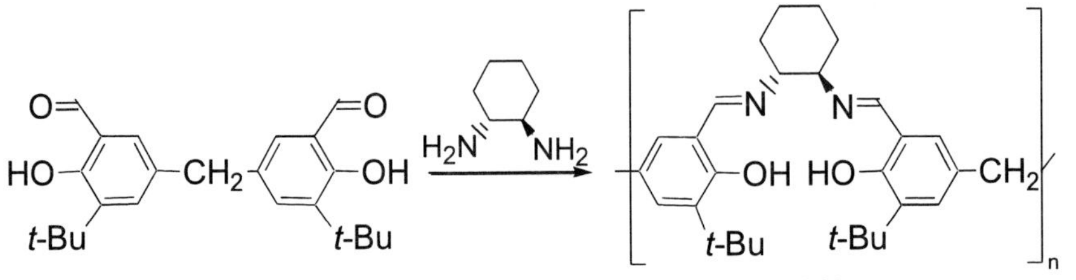

340

Scheme 145.

Other **340**-Co-salen complexes containing Lewis acid were studied [214] (Scheme 146).

Monomeric salen Co(II) or Co(III) unit are supposed to be attached to the polymer salen catalyst by means of the Lewis acid. Thus, salen units were linked together as a dimeric form (polysalen **346** and **348**, Scheme 145). It was observed a great increase of the catalytic activity for the HKR reaction of various epoxides. For example, in the presence of 0.4 mol% polysalen catalysts **346** and **348** the HKR reaction of epichlorhydrine led to about 98% ee, after 6h whereas with complex **345** ee was up to 55%. As dissociation of monomeric salens from the polymer backbone partially occurred after their first use, the polymeric salen **346** and **348** lost their catalytic activity. It is noteworthy that the activity of the salen catalyst could be recovered upon treatment with monomeric salen.

Contrary to Mn salen **256**$_{Mn}$ (Scheme 126), which is used for the AE, Laibinis used the heterogeneous Cr salen **256**$_{Cr}$ catalyzed ring epoxide opening in the presence of TMSN$_3$ [190]. In the conditions of ring opening epoxide, ee reach 36% and yield 47% for the propylene oxide. But the catalytic system is stable and could be reused 3 times without loss of activity and enantioselectivity.

Weck [203] has performed the HKR of several epoxides by means of polymer-supported cobalt salen catalysts containing different counterions (**325**/Co-OAc, **325**/Co-I, **325**/Co-OTs, **325**/Co Scheme 137). **325**/Co-I and **325**/Co-OTs catalysts have shown higher activities than **325**/Co-OAc and ee could be up to > 99% with a conversion of 55% after 1h with **325**/Co-I for the HKR of epichlorhydrine. These catalysts can only be recycled once as a result of their decreased solubility after the reoxidation step.

Weck [215] has also described polystyrene-supported chiral salen ligands **349** and **350** synthesized respectively by radical homopolymerization or copolymerization of an unsymmetrical styryl-substituted salen monomer (Scheme 147). They were easily converted into their Co(III) acetate complexes and then tested in the HKR of epichlorhydrine at room temperature. The reaction was carried out with 0.5 mol% of the catalyst and after 1h, conversion is about 55% and ee >98% even after four recycling (catalyst **350b**).

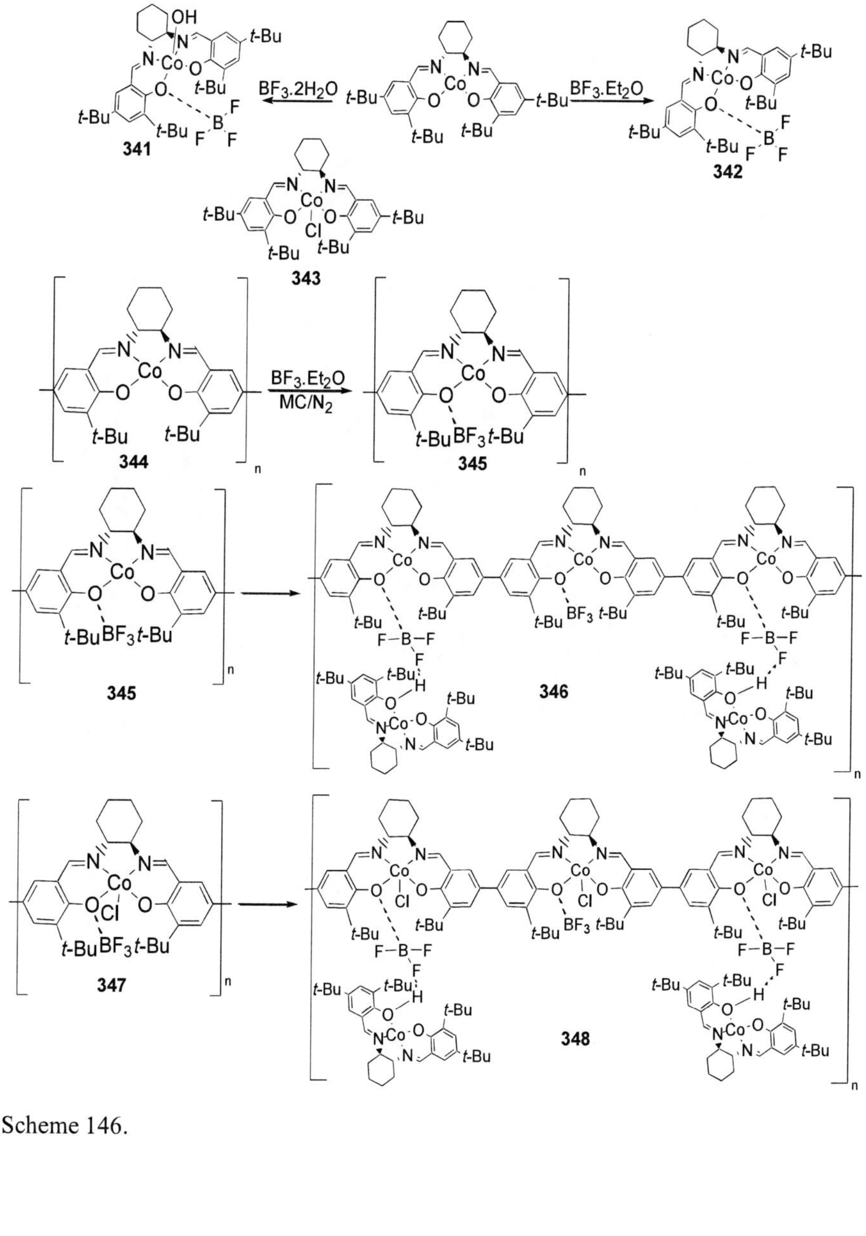

Scheme 146.

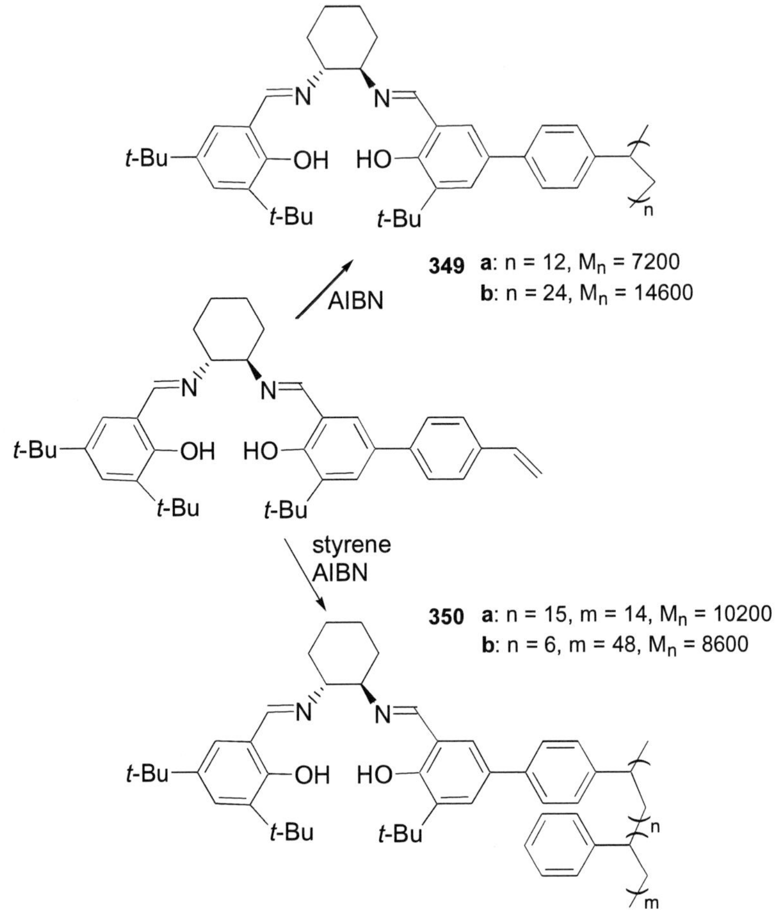

349 **a**: n = 12, M$_n$ = 7200
      b: n = 24, M$_n$ = 14600

350 **a**: n = 15, m = 14, M$_n$ = 10200
      b: n = 6, m = 48, M$_n$ = 8600

Scheme 147.

Polyethylene glycol–salen-Co-OAc (Scheme 140) were applied in the HKR of several terminal epoxides with a catalyst loading of 0.5 mol% using 0.55 equiv. of water in THF. A positive dendritic effect was observed; the dendritic backbone was supposed to increase the intramolecular cooperation between Co active site by bringing the reactive site of the reactant adjacent to each other (cooperative bimetallic catalysis) [208]. This reason was evoked to explain that better results were obtained with the polymeric catalyst contrary to the non polymeric one. With epichlorhydrine, the conversion reached 52.5% and the enantioselectivity 90.5%.

Chapter 10

MISCELLANEOUS

Few other asymmetric reactions have been performed using insoluble or soluble polymer-supported ligands. The first example is a Mukaiyama-aldol condensation between silyl ketene acetal and different aldehydes using polymeric Box analog of **99** (Scheme 46) as chiral ligands and $Cu(OTf)_2$ as metal source in water (Scheme 148) [216]. When using benzaldehyde as substrate, yields were very low (12-34%) and ee were moderate (40-62%) whater the polymer-supported Box. The same level of enantioselectivity was observed with other aldehydes while the yield was better with all the ligand/Cu complexes used.

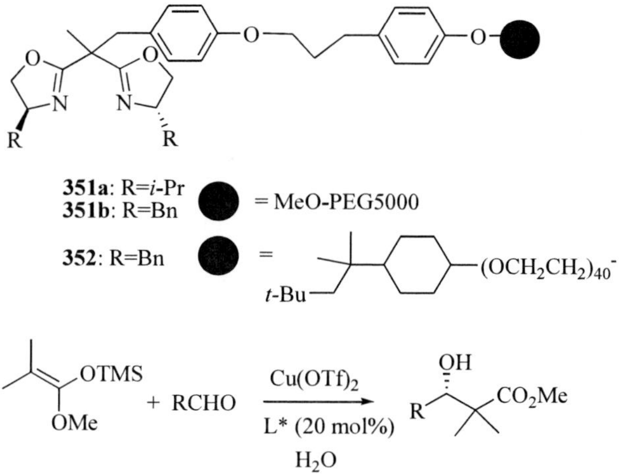

351a: R=*i*-Pr
351b: R=Bn   ● = MeO-PEG5000

352: R=Bn

Scheme 148.

Insoluble polymer-supported Box **82** (n=1) (Scheme 40) was also used in an enantioselective glyoxylate-ene reaction (Scheme 149) [217]. In most cases, although the yields and the ee were satisfactory, they were slightly slower than those observed with the corresponding non-polymeric Box.

The possibility to recycle the catalyst was examined in the ene reaction of α-methylstyrene and ethyl glyoxylate. The conversion dropped from 90 to 60% after the fourth recycle (conversion was evaluated after six hours for each run) while the enantioselectivity remained constant.

Furthermore, a continuous flow-system has been developed using the polymer **82** and overall 23 mmol of α-methylstyrene have been transformed into the corresponding product after 5 consecutive batches. The enantio-selectivity did not change between the successive runs (close to 90% ee). The TTN was estimated at 44 or 51 versus the supported Box and copper salt respectively.

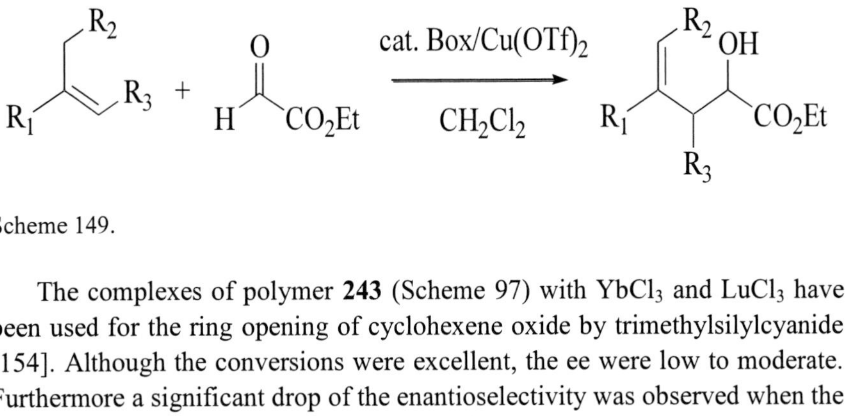

Scheme 149.

The complexes of polymer **243** (Scheme 97) with $YbCl_3$ and $LuCl_3$ have been used for the ring opening of cyclohexene oxide by trimethylsilylcyanide [154]. Although the conversions were excellent, the ee were low to moderate. Furthermore a significant drop of the enantioselectivity was observed when the recovered supported Yb-complex was reused.

The same complexes were then evaluated for the enantioselective TMSCN addition to benzaldehyde. For this reaction, both yield and ee were better with the Lu-complex.

The asymmetric TMSCN addition to benzaldehyde was also performed in a microreactor using the polymeric catalyst **353** (Scheme 150) [218]. A normal batch procedure was first carried out and the protected cyanohydrin was obtained with 91% conversion and 73% ee. When a microreactor using catalyst was employed, same result was observed with a flow rate of 0.8µL.

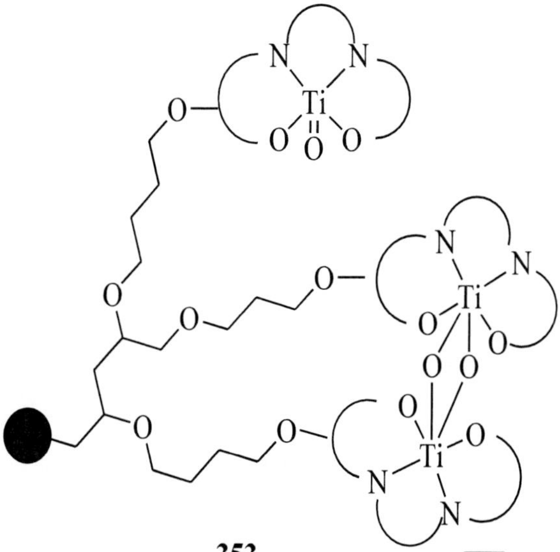

353

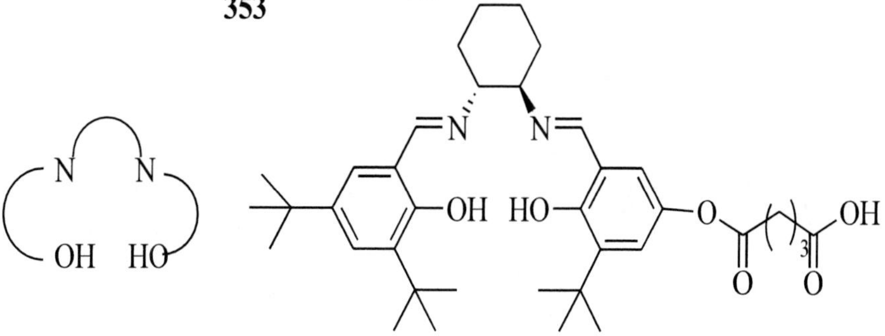

Scheme 150.

Enantioselective sulfoxidation has been also studied using polymeric aminoalcohol-derived Schiff bases. Polymers **354-357** have been prepared by copolymeristation of the corresponding monomer with either MMA and EGDMA or styrene and DVB (Scheme 151) [219]. These four polymers were then stirred with VO(acac)$_2$ for 4h at rt to lead to the vanadium complexes. For the enantioselective oxidation of thioanisole, although the yields were similar to those obtained with the homogeneous analogs, the enantioselectivity were lower. The best selectivity was obtained with the catalyst derived from **355**.

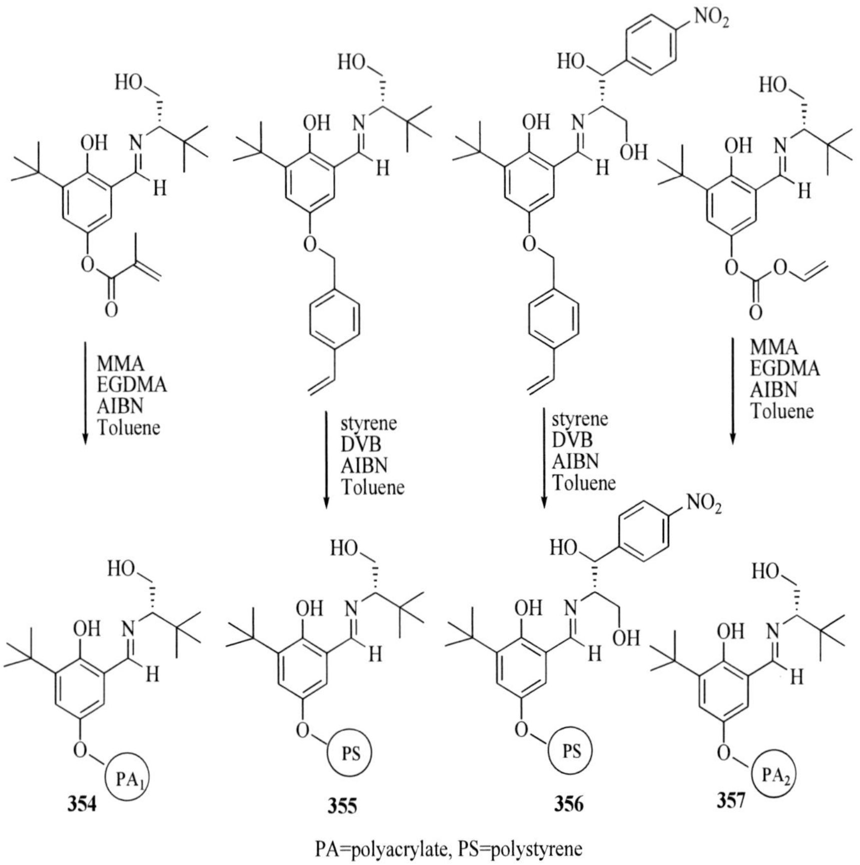
MMA
EGDMA
AIBN
Toluene
styrene
DVB
AIBN
Toluene
styrene
DVB
AIBN
Toluene
MMA
EGDMA
AIBN
Toluene
354
355
356
357
PA=polyacrylate, PS=polystyrene

Scheme 151.

Chapter 11

CONCLUSION

The performance of the soluble catalysts supported on a polymer or of the heterogeneous catalysts is highly dependent on the polymer morphology which is, in the last case, mainly controlled by the degree of crosslinking, the nature of the crosslinker and nature of the porogen.

For soluble polymer catalysts, the catalytic performances are generally close to that observed with homogeneous non-supported catalysts. This behaviour is mainly inherent to high mobility of the bound catalyst and a good mass transport. However, the stability of the catalyst depends on the reaction conditions, some catalytic systems being not stable in some solvents. Their recovery needs filtrations or precipitations, sometimes under inert atmosphere. But, their use could be interesting for an industrial point of view, particularly in continuous flow systems involving membranes where important conversions of the substrate with good selectivity could be involved.

As for insoluble catalysts they could be separated into two series.The ligand could be anchored onto swellable or unswellable polymers.

In the first case, the support is formed by a crosslinked polymer, mostly a polystyrene crosslinked with 0.5 to 3% DVB.

In the second case, pratically unswellable polymers formed the support. This latter is a highly crosslinked polymer. A large variety of solvent could be used without changing the texture of the catalyst. In order to obtain a good mass transport, it is generally necessary to obtain immobilized catalysts with a pore size enough larger to include the size of the metal complex catalyst and the substrate. This parameter depends on the nature of the porogen. These immobilized catalysts must present a large surface area

For the specific molecular imprinting polymer, the catalyst is efficient if a good mass transport is possible and particularly if the chiral cavity fits perfectly with the size and the structure of the substrate. It is the reason why in most of the cases the shape of the polymer is specific to a single substrate.

Reaction performed with monolithic polymer-supported catalysts, lead generally to good to excellent behaviour in terms of catalytic activity and enantioselectivity and as the reaction is performed without stirring, the catalytic system is generally very stable.

The catalyst ability for recycling and reuse is in correlation to the stability of the catalytic system itself. A simple change of the nature of the polymer support could increase the stabilization and thus the recyclability of the catalytic system.

Concerning the nature of the nitrogen-containing ligand itself, immobilized derivatives of DPEN revealed to be the most efficient in the reduction of prochiral ketones by catalytic hydrogenation in the presence of BINAP or by HTR and CBS-supported ligands for the enantioselective hydride reduction.

As for C-C bond formation *via* an addition of organometallic reagent to aldehyde, ketone or imine or allylation, supported-complexes of prolinol, ephedrine and oxazoline derivated constitute the best catalytic systems.

Most of the polymeric-catalytic systems used for the dihydroxylation or aminohydroxylation of C=C bond is issued from alkaloid, mainly cinchona which has been proved to be the best ligand for this type of reaction.

For the epoxidation or kinetic resolution of terminal epoxide, polymer-supported salens have shown to be either as efficient as or more efficient than their non-supported homologs.

It is noteworthy that concerning the loading, low loading in ligand and metal are required to achieve asymmetric hydrogenation, C-C bond formation or dihydroxylation or aminohydroxylation. But, high loading is necessary for epoxidation or resolution of terminal epoxides if salen derivatives ligands are employed.

To conclude, the nitrogen-containing ligands are easy to handle, to synthesize and to support them on polymers, thus this constitutes a crucial advantage for using them in various separation systems (precipitation, ultrafiltration, membrane reactor, biphasic solid/liquid or liquid/liquid).

REFERENCES

[1] *Chirality in Industry: The Commercial Manufacture and Applications of Optically Active Compounds* (Eds.: Collins A. N., Sheldrake G. N., Croby J.), Wiley, 1995; b) Sheldon R. A., Chirotechnology, Dekker, New York, 1993.

[2] Ramos Tombo G. M., Bellus D. *Angew. Chem. Int. Ed.* 1991, *30*, 1193-1215.

[3] Noyori R., *CHEMTECH* 1992, *22*, 360-367

[4] a) R. A. Sheldon, *CHEMTECH* 1994, *March*, 38-47; b) R. A. Sheldon, *Chem. Ind.* 1992, 903-906; c) R. A. Sheldon *Chem. Ind.* 1997, 12-15.

[5] Kragl U., Dwars T., *Trends biotechnol.* 2001, *19*, 442-449.

[6] D. J. Bayston, M. E. C. Polywka in *Chiral Catalyst Immobilization and recycling,* (Eds.: De Vos D. E., Vankelecom I. F. J., Jacobs P. A) Wiley-VCH, Weinheim, 2000, pp. 211-234.

[7] Wulff G., *Angew; Chem; Int. Ed.* 1995, *34*, 1812-1832

[8] (a) Ohkuma, T.; Kitamura, M.; Noyori, R. *In Catalytic asymmetric synthesis,* Editor: Ojima I. 2nd Ed. Wiley-VCH: New York, NY, 2000, pp. 1-110. (b) Noyori, R.; Ohkuma, T. *Angew. Chem. Int. Ed.* 2001, 40-73.

[9] Ohkuma, T.; Ooka, H; Hashiguchi, S; Ikariya, T; Noyori, R.; *J. Am. Chem. Soc.* 1995, *117*, 2675-2676.

[10] Ohkuma, T.; Koizumi M.; Doucet, H.; Pham, T.; Kozawa, M.; Murata, K.; Katayama, E.; Yokozawa, T. Ikariya, T; Noyori, R.; *J. Am. Chem. Soc.* 1998, *120*, 13529-13530].

[11] Ohkuma T., Takeno H., Honda Y., Noyori R., *Adv. Synth. Catal.* 2001, 369-375.

[12] Itsuno, S.; Tsuji, A.; Takahashi, M. *Tetrahedron Lett.* 2003, *44*, 3825-3828.

[13] (a) Takahashi, M.; Haraguchi, N.; Itsuno, S. *Tetrahedron: Asymmetry* 2008, *19*, 60-66. (b) Haraguchi, N.; Prasad, N.; Uchiyama T.; Itsuno, S. *Macromol. Synop.* 2007, *254*, 67-73.

[14] a) Itsuno, S.; Tsuji, A.; Takahashi, M. *J. Polym. Sci., Part A: Polym. Chem.* 2004, *42*, 4556-4562. b) Takahashi, M.; Tsuji, A.; Chiba, M.; Itsuno, S. *React. Funct. Polym.* 2005, *65*, 1-8.

[15] (a) Itsuno, S.; Chiba, M.; Takahashi, M.; Arakawa, Y.; Haraguchi, N. *J. Organomet. Chem.* 2007, *692*, 487-494.(b) Haraguchi, N.; Chiba, M.; Takahashi, M.; Itsuno, S.; *Macromol. Synop.* 2007, *249-250*, 365-372.

[16] Chiwara, V. I.; Haraguchi, N.; Itsuno, S. *J. Org. Chem.,* 2009, *74*, 1391-1393

[17] a) Johnstone, R. A. W.; Wilby, A. H.; Entwistle I. D. *Chem. Rev.* 1985, *85*, 129-170. b) Zassinovich, G.; Mestroni, G. *Chem. Rev.* 1992, *92*, 1051-1069.

[18] Gamez, P., Fache, F.; Mangeney, P.; Lemaire, M. *Tetrahedron Lett.* 1993, *34*, 6897-6898

[19] Touchard F.; Gamez, P.; Fache, F.; Lemaire, M. *Tetrahedron Lett.* 1997, *38*, 2275-2278

[20] (a) Dunjic, B.; Gamez, P.; Fache, F.; Lemaire, M. *J. Appl. Polym. Sci.* 1996, *59*, 1255-1262. (b) Gamez, P.; Dunjic, B.; Fache, F.; Lemaire, M. *J. Chem. Soc., Chem. Comm.* 1994, 1417-1418. (c) Lemaire, M. *Pure Appl. Chem.* 2004, *76*, 679-688.

[21] Touchard F.; Fache, F.; Lemaire, M. *Eur. J. Org. Chem.* 2000, 3787-3792.

[22] (a) F. Touchard, P. Gamez, F. Fache, M. Lemaire, *Tetrahedron Lett.* 1997, *38*, 2275-2278. (b) F. Touchard, M. Bernard, F. Fache, M. Lemaire, *J. Mol. Catal. A: Chem.* 1999, *140*, 1-11.

[23] Hashiguchi, S.; Fujii, A.; Takehara, J.; Ikariya, T.; Noyori, R. J. *J. Am. Chem. Soc.* 1995, *117*, 7562-7563

[24] ter Halle, R.; Schulz, E.; Lemaire, M. *Synlett*, 1997, 1257-1258.

[25] Arakawa, Y.; Haraguchi, N.; Itsuno, S. *Tetrahedron Lett.* 2006, *47*, 3239-3243.

[26] Arakawa, Y.; Chiba, A.; Haraguchi, N.; Itsuno, S. Adv. Synth. Catal. 2008, 350, 2295-2304.

[27] Bayston, D. J.; Travers, C. B.; Polywka, E. C. *Tetrahedron: Asymmetry* 1998, *9*, 2015-2018.

[28] Li, Y.; Li, Z.; Li, F.; Wang, Q.; Tao, F. *Org. Biomol. Chem.* 2005, *3*, 2513-2518.

[29] Li, X.; Chen, W.; Hems, W.; King, F.; Xiao, J. *Tetrahedron Lett.* 2004, *45*, 951-953.

[30] Li, X.; Wu, X.; Chen, W.; Hancock, F. E.; King, F.; Xiao, J. *Org. Lett.* 2004, *6*, 3321-3324.

[31] Liu, J.; Zhou, Y.; Wu, Y.; Li, X.; Chan, A. S. C. *Tetrahedron: Asymmetry* 2008, *19*, 832-837.

[32] Gamez P., Dunjic B., Pinel C., Lemaire M., *Tetrahedron Lett.*, 1995 *36*, 8779-8783.

[33] Locatelli, F.; Gamez, P.; Lemaire, M. *Stud. Surf. Sci. Catal.* 1997, *108* (Heterogeneous Catalysis and Fine Chemicals IV), 517-522.

[34] Locatelli F., Gamez P., Lemaire M., *J. Mol. Catal. A. Chem.* 1998, *135*, 89-98.

[35] Polborn K., Severin K., *Chem. Commun.* 1999, 2481-2482.

[36] Polborn K., Severin K., *Chem. Eur. J.* 2000, *6*, 4604-4611.

[37] Polborn K., Severin K., *Eur. J. Inorg.* 2000, 1687-1692.

[38] (a) W. E. Silverthorn, *Adv. Organomet. Chem.* 1975, *13*, 47-137. (b) E. L. Muetterties, J. R. Bleeke, E. J. Wucherer, T. Albright, *Chem. Rev.* 1982, *82*, 499-525. c() Takehara, J.; Hashiguchi S., Fujii A., Inoue S., Ikariya T., Noyori R. *Chem. Commun.*, 1996, 233 – 234.

[39] Rolland, A.; Hérault, D.; Touchard, F.; Saluzzo, C.; Duval, R.; Lemaire, M. *Tetrahedron: Asymmetry* 2001, *12*, 811-815.

[40] Hérault, D.; Saluzzo, C.; Lemaire, M. *Reactive and functional polymers,* 2006, *66*, 567-577.

[41] Hérault, D.; Saluzzo, C.; Lemaire, M. *Tetrahedron: Asymmetry* 2006, *17*, 1944-1951.

[42] Saluzzo, C.; Lamouille, T.; Hérault, D.; Lemaire, M. *Bioorg. Med. Chem. Lett.* 2002, *12*, 1841-1844.

[43] Bastin, S.; Eaves, R.J.; Edwards, C.W.; Ichihara, O.; Whittaker, M.; Wills, M. *J. Org. Chem.* 2004, *69*, 5405-5412.

[44] Sanda, F.; Araki, H.; Masuda, T. *Chem. Lett.* 2005, *34*, 1642-1643.

[45] Breysse, E., Pinel, C; Lemaire, M. *Tetrahedron: Asymmetry* 1998, *9*, 897-900.

[46] Hirao, A.; Itsuno, S.; Nakahama, S. Yamasaki, N. *J. Chem. Soc, Chem. Commun.* 1981, 315-317.

[47] a) Corey, E. J.; Bakashi, R. K.; Shibata, S. *J. Am. Chem. Soc.* 1987, *109*, 5551-5553. b) Corey, E. J.; Bakashi, R. K.; Shibata, S.; Chen, C. P.; Singh, V. K. *J. Am. Chem. Soc.* 1987, *107*, 7925-7926.

[48] (a) Deloux, L.; Srebnik, M. *Chem. Rev.* 1993, *93*, 763-784. (b) Wallbaum, S.; Martens, J. *Tetrahedron: Asymmetry* 1992, *3*, 1475-1504. (c) Singh, V. K. *Synthesis*, 1992, 605-617. d) Corey, E. J. *Pure Appl. Chem.* 1990, *7*, 1209-1216.

[49] (a) Lecavalier, P.; Bald, E.; Jiang, Y.; Fréchet, J.M.J.; Hodge, P. *React. Polym.* 1985, *3*, 315-326. (b) Altava, B.; Burguete, M.I.; Collado, M.; Garcia-Verdugo, E.; Luis, S.V.; Salvador, R.J.; Vicent, M. J. *Tetrahedron Lett.* 2001, *42*, 1673-1675.

[50] Itsuno, S.; Ito, K.; Hirao, A.; Nakahama, S. *J. Chem. Soc., Perkin Trans. 1* 1984, 2887-2893.

[51] (a) Itsuno, S.; Nakano, M.; Ito, K.; Hirao, A.; Owa, M.; Kanda, N; Nakahama, S. *J. Chem. Soc., Perkin Trans. 1* 1985, 2615-2619 (b) Itsuno, S.; Sakurai, Y.; Shimizu, K.; Ito, K. *J. Chem. Soc; Perkin Trans I* 1990, 1859-1863.

[52] Adjidjonou, K.; Caze, C. *Eur. Polym. J.* 1995, *31*, 749-754.

[53] Itsuno, S.; Ito, K.; Maruyama, T.; Kanda, N; Hirao, A.; Nakahama, S. *Bull. Chem. Soc. Jpn.* 1986, 3329-3331.

[54] Giffels, G.; Beliczey, J.; Felder, M.; Kragl, U. *Tetrahedron: Asymmetry* 1998, *9*, 691-696.

[55] (a) Blanton, J. R. *React. Funct. Polym.* 1997, *33*, 61-69. (b) Caze, C.; El Moualij, N.; Hodge, P.; Lock, C. J.; Ma, J. *J. Chem. Soc., Perkin Trans. 1* 1995, 346-349. (c) Franot, C.; Stone, G. B.; Engeli, P.; Spöndlin, C.; Waldvogel, E. *Tetrahedron: Asymmetry* 1995, *6*, 2755-2766.

[56] Price, M.D.; Sui, J.K.; Kurth, M.J.; Shore, N.E. *J. Org. Chem.* 2002, *67*, 8086-8089.

[57] Varela, M.C.; Dixon, S. M.; Price, M. D.; Merit, J. E.; Berget, P. E.; Shiraki, S.; Kurth, M. J.; Schore, N. E. *Tetrahedron* 2007, *63*, 3334-3339.

[58] Degni, S.; Wilén, C. E.; Rosling, A. *Tetrahedron: Asymmetry* 2004, *15*, 1495-1499.

[59] Kell, R. J.; Hodge, P.; Snedden, P.; Watson, D. *Org. Biomol. Chem.* 2003, *1*, 3238-3243.

[60] Hu, J.-B.; Zhao, G.; Yang, G.-S.; Ding, Z.-D. *J. Org. Chem.* 2001, *66*, 303-304.

[61] Wang, G.; Liu, X.; Zhao, G. *Tetrahedron: Asymmetry*, 2005, *16*, 1873-1879.

[62] Zhao, G.; Hu, J. B.; Qian, Z. S.; Yin, W. X. *Tetrahedron: Asymmetry*, 2002, *13*, 2095-2098.

[63] Chen, F., E; Yuan, J. L; Dai, H. F.; Kuang, Y. Y.; Chu, Y. *Synthesis* 2003, 2155-2160.

[64] Cha, R. T.; Li, Q. L.; Chen, N. L.; Wang, Y. P. *J. Appl. Polym. Sci.* 2007, *103*, 148-152.

[65] Enders, D.; Gielen, H.; Breuer K. *Molecules online* 1998, *2*, 105-108.

[66] For example: Itsuno, S.; Ito, K; Maruyama, T.; Kanda, N.; Hirao, A.; Nakahama, S. *Bull. Chem. Soc. Jpn* 1987, *60*, 395-396.

[67] Itsuno, S.; Nakano, M.; Ito, K; Hirao, A.; Owa, M.; Kanda, N.; Nakahama, *J. Chem. Soc., Perkin Trans I*, 1985, 2615-2619.

[68] Itsuno, S.; Matsumoto, T.; Sato, D.; Inoue, T. *J. Org. Chem.* 2000, *65*, 5879-5881.

[69] Malkov, A. V.; Figlus, M.; Kočovsky, P. *J. Org. Chem.* 2008, *73*, 3985-3995.

[70] Haraguchi, N.; Tsuru, K.; Arakawa, Y.; Itsuno, S. *Org. Biomol. Chem.* 2009, *7*, 69-75.

[71] (a) Salaün, J. *Chem. Rev.* 1989, *89*, 1247–1270 (b) Pellissier, H. *Tetrahedron* 2008, *64*, 7041-7095.

[72] Nozaki, H.; Moriuti, S.; Takaya, H.; Noyori, R. *Tetrahedron Lett.* 1966, 43, 5239-5244.

[73] Doyle M. P. *In Catalytic Asymmetric Synthesis;* Editor: Ojima I. 2nd Ed. Wiley-VCH: New York, NY, 2000, pp. 191-228.

[74] (a) Lowenthal, R. E.; Abiko, A.; Masamune, S. *Tetrahedron Lett.* 1990, *31*, 6005-6008. (b) Lowenthal, R. E.; Masamune, S. *Tetrahedron Lett.* 1991, *32*, 7373-7376.

[75] (a) Desimoni, G.; Faita, G.; and Jørgensen, K. A. *Chem. Rev.* 2006, *106*, 3561-3651. (b) Lebel, H.; Marcoux, J-F.; Molinaro C.; Charrette, A. B. *Chem. Rev.* 2003, *103*, 977–1050. (c) Desimoni, G.; Faita, G.; Quadrelli, P. *Chem. Rev.* 2003, *103*, 3119–3154.

[76] Carmona, A.; Corma, A.; Iglesias, M.; Sanchez, F. *Inorg. Chim. Acta* 1996, *244*, 79- 85.

[77] (a) Fraile, J. M.; Garcia, J. I.; Mayoral, J. A.; Tarnai, T *Tetrahedron: Asymmetry* 1997, *8*, 2089- 2092. (b) Fraile, J. M.; Garcia, J. I.; Mayoral, J. A.; Tarnai, T *Tetrahedron: Asymmetry* 1998, *9*, 3997-4008.

[78] Burguete, M. I.; Fraile, J. M.; Garcia, J. I.; Garcia-Verdugo, E.; Luis, S. V.; Mayoral, J.A. *Org. Lett.* 2000, *2*, 3905-3908.

[79] Doyle, M. P. *In Catalytic Asymmetric Synthesis;* Editor : Ojima I. 2nd Ed. Wiley-VCH: New York, NY, 2000, pp. 191-228.

[80] Burguete, M. I.; Fraile, J. M.; Garcia, J.I.; Garcia-Verdugo, E.; Herrerias, C. I.; Luis, S.V.; Mayoral, J.A. *J. Org. Chem.* 2001, *66*, 8893-8901.

[81] Burguete, M. I.; Diez-Barra, E.; Fraile, J. M.; Garcia, J. I.; Garcia-Verdugo,E.; González, R.; Herrerias, C. I.; Luis, S.V.; Mayoral, J. A. *Bioorg. Med. Chem. Lett.* 2002, *12*, 1821-1824.

[82] Diez-Barra, E.; Fraile, J. M.; Garcia, J. I.; Garcia-Verdugo, E.; Herrerias, C. I.; Luis, S. V.; Mayoral, J. A; Sánchez-Verdu, P. Tolosa, J. *Tetrahedron: Asymmetry* 2003, *14*, 773-778.

[83] Burguete, M. I.; Fraile, J. M.; Garcia-Verdugo, E.; Luis, S. V.; Martinez-Merino, V.; Mayoral, J. A. *Ind. Eng. Chem. Res.* 2005, *44*, 8580-8587.

[84] Mandoli, A; Orlandi, S; Pini, D, Salvadori, P. *Chem. Comm.* 2003, 2466-2467.

[85] Mandoli, A.; Garzelli, R.; Orlandi, S.; Pini, D.; Lessi, M.; Salvadori, P. *Catal. Today* 2009, *140*, 51-57

[86] Yifei, H. Chemical Journal on Internet, 2004, 6, 79-83 Publisher: Chemical Journal on Internet, http://www.chemistrymag.org/cji/2004/06b079ne.htm.

[87] Knight, J.G.; Belcher, P.E. *Tetrahedron: Asymmetry* 2005, *16*, 1415-1418.

[88] Werner, H.; Herrerías, C. I.; Glos, M.; Gissibl, A.; Fraile, J. M.; Pérez, I.; Mayoral, J. A.; Reiser, O. *Adv. Synth. Catal.* 2006, *348*, 125-132.

[89] Cornejo, A.; Fraile, J. M.; Garcia, J. I.; Garcia-Verdugo, E.; Gil, M. J.; Legaretta, G.; Luis, S. V.; Martinez-Merino, V.; Mayoral, J. A. *Org. Lett.* 2002, *4*, 3927-3930.

[90] Cornejo, A.; Fraile, J. M.; Garcia, J. I.; Gil, M. J.; Luis, S. V.; Martinez-Merino, V.; Mayoral, J. A. *J. Org. Chem.* 2005, *70*, 5536-5544.

[91] Burguete, M. I.; Cornejo, A.; Garcia-Verdugo, E.; Gil, M. J.; Luis, S. V.; Mayoral, J. A.; Martinez-Merino, V.; Sokolova, M. *J. Org. Chem.* 2007, *72*, 4344-4350.

[92] Donzé, C.; Pinel, C.; Gazellot, P, Taylor, P. L. *Adv. Synth. Catal.* 2002, *344*, 906-910.

[93] Cornejo, A.; Martinez-Merino, V.; Gil, M.J.; Valerio, C.; Pinel, C. *Chem. Lett.* 2006, *35*, 44-45.

[94] Cornejo, A.; Fraile, J. M.; Garcia, J. I.; Gil, M. J.; Martinez-Merino, V.; Mayoral, J. A. *Mol. Diversity* 2003, *6,* 93-105.

[95] Ghosh, A. K.; Mathivanan, P.; Cappiello, J. *Tetrahedron: Asymmetry*, 1998, *9,* 1-45.

[96] Annunziata, R.; Benaglia, M.; Cinquini, M.; Cozzi, F.; Pitillo, M. *J. Org. Chem.* 2001, *66*, 3160-3166.

[97] Glos, M.; Reiser, O. *Org. Lett.* 2000, *2*, 2045-2048.

[98] Ferrand, Y. ; Le Maux, P.; Simonneaux, G. *Tetrahedron: Asymmetry* 2005, *16*, 3829–3836.

[99] Ferrand, Y. ; Poriel, C. ; Le Maux, P.; Rault-Berthelot, J. ; Simonneaux, G. *Tetrahedron: Asymmetry* 2005, *16*, 1463-1472.

[100] Doyle, M. P.; Timmons, D. J.; Tumonis, J. S.; Gau, H.-M.; Blossey, E. C. *Organometallics* 2002, *21*, 1747-1749.

[101] Jorgensen, K. A. *Angew. Chem. Int. Ed.* 2000, *39*, 3558-3688.

[102] Hallman, K; Moberg, C *Tetrahedron: Asymmetry*, 2001, *12*, 1475-1478.

[103] Rechavi, D; Lemaire, M. *J. Mol. Catal. A: Chem.* 2002, *182*, 239-247.

[104] Itsuno, S.; Kamahori, K.; Watanabe, K.; Koizumu, T.; Ito, K. *Tetrahedron: Asymmetry* 1994, *5*, 523-526.

[105] Itsuno, S.; Watanabe, K.; Koizumi, T.; Ito, K. *React. Polym.* 1995, *24*, 219-227.

[106] Kamahori, K.; Tada S.; Ito, K. Itsuno, S.; *Tetrahedron: Asymmetry* 1995, *6*, 2547-2555.

[107] Kamahori, K.; Ito, K. Itsuno, S.; *J. Org. Chem.* 1996, *61*, 8321-8324.

[108] Sellner, H.; Karjalainen, J. K.; Seebach, D. Chem. Eur. J. 2001, 7, 2873-2887.

[109] Mellah, M.; Ansel, B.; Patureau, F.; Voituriez, A.; Schulz, E. *J. Mol. Catal., A: Chem.* 2007, *272*, 20-25.

[110] Zulauf, A.; Mellah, M.; Guillot, R.; Schulz, E. *Eur. J. Org. Chem.* 2008, 2118-2129.

[111] Noyori, R.*Asymmetric Catalysis in Organic Synthesis*, Wiley, NewY ork, 1994, pp. 225-297.

[112] (a) Soai, K.; Niwa, S. *Chem. Rev.* 1992, *92*, 833-856. (b) Pu, L.; Yu, H.-B. *Chem. Rev.* 2001, *101*, 757-824.

[113] Itsuno, S.; Fréchet, J. M. J. *J. Org. Chem.* 1987, *52*, 4140-4142.

[114] Soai, K.; Niwa, S.; Watanabe, M. *J. Org. Chem.* 1988, *53*, 927-928.

[115] Soai, K.; Niwa, S.; Watanabe, M. *J. Chem. Soc., Perkin Trans. 1* 1989, 109-113.

[116] Soai, K.; Suzuki, T.; Shono, T. *J. Chem. Soc., Chem. Commun.* 1994, 317-318.

[117] Itsuno, S.; Sakurai, Y.; Ito, K.; Maruyama, T.; Nakahama, S.; Fréchet, J.M.J. *J. Org. Chem.* 1990, *55*, 304-310.

[118] Watanabe, M.; Soai, K. *J. Chem. Soc., Perkin Trans. 1* 1994, 837-842.

[119] Suzuki, T.; Shibata, T.; Soai, K. *J. Chem. Soc., Perkin Trans. 1* 1997, 2757-2760.

[120] Suzuki, T.; Narisada, N.; Shibata, T.; Soai, K. *Tetrahedron: Asymmetry* 1996, *7*, 2519-2522.

[121] Hosoya, K.; Tsuji, S.; Yoshizako, K.; Kimata, K.; Araki, K.; Tanaka, N. *React. Funct. Polym.* 1996, *29*, 159-166.

[122] Vidal-Ferran, A.; Bampos, N.; Moyano, A.; Pericàs, M. A.; Riera, A.; Sanders, J. K. M. *J. Org. Chem.* 1998, *63*, 6309-6318.

[123] (a) Vidal-Ferran, A.; Moyano, A.; Pericàs, M. A.; Riera, A. *J. Org. Chem.* 1997, *62*, 4970-4982. (b) Vidal-Ferran, A.; Moyano, A.; Pericàs, M. A.; Riera, A. *Tetrahedron Lett.* 1997, *38*, 8773-8776.

[124] Sung, D. W. L.; Hodge, P.; Stratford, P. W. *J. Chem. Soc., Perkin Trans. 1* 1999, 1463-1472.

[125] Hodge, P.; Sung, D. W. L.; Stratford, P. W. *J. Chem. Soc., Perkin Trans. 1* 1999, 2335-2342.

[126] ten Holte, P.; Wijgergangs, J.-P.; Thijs, L.; Zwanenburg, B. *Org. Lett.* 1999, *1*, 1095-1097.

[127] Degni, S.; Wilén, C.-E.; Leino, R. *Org. Lett.* 2001, *3*, 2551-2554.

[128] Degni, S.; Wilén, C.-E.; Leino, R. *Tetrahedron: Asymmetry* 2004, *15*, 231-237.

[129] Hu, Q.-S.; Sun, C.; Monaghan, C. E. *Tetrahedron Lett.* 2001, *42*, 7725-7728.

[130] Burguete, M. I.; Garcia-Verdugo, E.; Vicent, M. J.; Luis, S. V.; Pennemann, H.; Graf von Keyserling, N.; Martens, J. *Org. Lett.* 2002, *4*, 3947-3950.

[131] Luis, S. V.; Altava, B.; Burguete, M. I.; Collado, M.; Escorihuela, J.; Garcia-Verdugo, E.; Vicent, M. J.; Martens, J. *Ind. Eng. Chem. Res.* 2003, *42*, 5977-5982.

[132] Burguete, M. I.; Collado, M.; Garcia-Verdugo, E.; Vicent, M. J.; Luis, S. V.; Graf von Keyserling, N.; Martens, J. *Tetrahedron* 2003, *59*, 1797-1804.

[133] Lesma, G.; Danieli, B.; Pasarella, D.; Sachetti, A.; Silvani, A. *Tetrahedron: Asymmetry* 2003, *14*, 2453-2458.

[134] Pericàs, M. A.; Castellnou, D.; Rodriguez, I.; Riera, A.; Solà, L. *Adv. Synth. Catal.* 2003, *345*, 1305-1313.

[135] Castellnou, D.; Fontes, M.; Jimeno, C.; Font, D.; Solà, L.; Verdaguer, X.; Pericàs, M. A. *Tetrahedron* 2005, *61*, 12111-12120.

[136] Castellnou, D.; Sol, L.; Jimeno, C.; Fraile, J. M.; Mayoral, J. A.; Riera, A.; Pericàs, M. A. *J. Org. Chem.* 2005, *70*, 433-438.

[137] Bastero, A.; Font, D. Pericàs, M.A. *J. Org. Chem.* 2007, *72*, 2460-2468.

[138] Wang, S.-X.; Chen, F.-E. *Chem. Pharm. Bull.* 2007, *55*, 1011-1013.

[139] Chai, L.-T.; Wang, Q.-R.; Tao, F.-G. *J. Mol. Catal. A: Chem.* 2007, *276*, 137-142.

[140] Kelsen, V.; Pierrat, P.; Gros, P.C. *Tetrahedron* 2007, *63*, 10693-10697

[141] El-Shehawy, A.A. *Tetrahedron* 2007, *63*, 5490-5500.

[142] El-Shehawy, A.A. *Tetrahedron* 2007, *63*, 11754-11762.

[143] El-Shehawy, A.A.; Sugiyama, K.; Hirao, A. *Tetrahedron: Asymmetry* 2008, *19*, 425-434.

[144] Page, P. C. B.; Allin, S. M.; Dilly, S. J.; Buckley, B. R. *Synth. Comm.* 2007, *37*, 2019-2030.

[145] Chen, C.; Hong, L.; Zhang, B.; Wang, R. *Tetrahedron: Asymmetry* 2008, *19*, 191-196.

[146] Itsuno, S.; El-Shehawy, A. A.; Sarhan, A. A. *React. Funct. Polym.* 1998, *37*, 283-287.

[147] Brouwer, A. J.; van der Linden, H. J.; Liskamp, R. M. J. *J. Org. Chem.* 2000, *65*, 1750-1757.

[148] Forrat, V. J.; Ramón, D. J. Yus, M. *Tetrahedron: Asymmetry* 2006, *17*, 2054-2058.

[149] Hui, X.-P.; Chen, C.-A.; Wu, K.-H.; Gau, H.M. *Chirality*, 2007, *19*, 10-15.

[150] Chen, C.-A.; Wu, K.-H.; Gau, H.M. *Polymer*, 2008, *49*, 1512-1519.

[151] (a) Gajare, A.S.; Shaikh, N.S.; Jnaneshwara, G. K.; Deshpande, V.H.; Ravindranathan, T.; Bedekar, A.V. *J. Chem. Soc., Perkin Trans. 1* 2000, 999-1001. (b) Shaikh, N.S.; Deshpande, V.H.; Bedekar, A.V. *Tetrahedron Lett.* 2002, *43*, 5587-5589.

[152] Bolm, C.; Hermanns, N.; Claßen, A.; Muñiz, K. *Bioorg. Med. Chem. Lett.* 2002, *12*, 1795-1798.

[153] Weissberg, A.; Halak, B.; Portnoy, M. *J. Org. Chem.* 2005, *70*, 4556-4559.

[154] Tilliet, M.; Lundgren, S.; Moberg, C.; Levacher, V. *Adv. Synth. Catal.* 2007, *349*, 2079-2084.

[155] Mao, J.; Bao, Z.; Guo, J.; Ji, S. *Tetrahedron* 2008, *64*, 9901-9905.

[156] Anyanwu, U. K.; Venkataraman, D. *Tetrahedron Lett.* 2003, *44*, 6445-6448.

[157] Jammi, S.; Rout, L.; Punniyamurthy, T. *Tetrahedron: Asymmetry* 2007, *18*, 2016-2020.

[158] Trost, B. M.; Van Vranken, D. L. *Chem. Rev.* 1996, *96*, 395- 422.

[159] Hallman, K.; Macedo, E.; Nordström, K.; Moberg, C. *Tetrahedron: Asymmetry* 1999 *10*, 4037-4046.

[160] Belda, O.; Lundgren, S.; Moberg, C. *Org. Lett.* 2003, *5*, 2275-2278.

[161] Bandini, M.; Benaglia, M.; Quinto, T.; Tommasi, S.; Umani-Ronchi, A. *Adv. Synth. Catal.* 2006, *348*, 1521-1527.

[162] Kolb, H. C.; Van Nieuwenhze, M. S.; Sharpless, K. B. *Chem. Rev.* 1994, *94*, 2483-2547.

[163] Sharpless, K. B.; Amberg, W.; Bennani, Y. L.; Crispino, G. A.; Hartung, J.; Jeong, K.-S.; Kwong, H.-L.; Morikawa, K.; Wang, Z.-M.; Xu, D.; Zhang, X.-L. *J. Org. Chem.* 1992, *57*, 2768-2771.

[164] Kim, B. M.; Sharpless, K. B. *Tetrahedron Lett.* 1990, *31*, 3003-3006.

[165] Pini, D.; Petri, A.; Nardi, A.; Rosini, C.; Salvadori, P. *Tetrahedron Lett.* 1991, *32*, 5175-5178.

[166] Lohray, B. B.; Thomas, A.; Chittari, P.; Ahuja, J. R.; Dhal, P. K. *Tetrahedron Lett.* 1992, *33*, 5453-5456.

[167] Pini, D.; Petri, A.; Salvadori, P. *Tetrahedron: Asymmetry* 1993, *49*, 2351-2354.

[168] Lohray, B.B.; Nandanan, E.; Bhushan, V. *Tetrahedron Lett.* 1994, *35*, 6559-6562.

[169] Rudolph, J.; Sennhenn, P. C.; Vlaar, C. P.; Sharpless, K. B. *Angew. Chem. Int. Ed. Engl.* 1996, *35*, 2810-2813.

[170] Nandanan, E.; Phukan, P.; Pais, G. C. G.; Sudalai, A. Ind. J. Chem. 1999, 38(B), 287-291.

[171] Pini, D.; Petri, A.; Salvadori, P. *Tetrahedron* 1994, *50*, 11321-11328.

[172] Petri, A.; Pini, D.; Salvadori, P. *Tetrahedron Lett.* 1995, *36*, 1549-1552.

[173] (a) Salvadori, P.; Pini, D.; Petri, A. *J. Am. Chem. Soc.* 1996, *119*, 6929-6930. (b) Salvadori, P.; Pini, D.; Petri, A. *Synlett* 1999, 1181-1190.

[174] Mandoli, A.; Pini, D.; Agostini, A.; Salvadori, P. *Tetrahedron: Asymmetry* 2000, *11*, 4039-4042.

[175] Song, C.E.; Roh, E.J.; Lee, S.; Kim, I. O *Tetrahedron: Asymmetry* 1995, *6*, 2687-2694.

[176] Song, C.E.; Yang, J.W.; Ha, H.J.; Lee, S. *Tetrahedron: Asymmetry* 1996, *7*, 645-648.

[177] Nandanan, E.; Sudalai, Ravindranathan, T. *Tetrahedron Lett.* 1997, *38*, 2577-2580.

[178] Mandoli, A.; Pini, D.; Fiori, M..; Salvadori, P. *Eur. J. Org. Chem.* 2005, 1271-1282.

[179] Han, H.; Janda, K. D. *J. Am. Chem. Soc.* 1996, *119*, 7632-7633.

[180] Han, H.; Janda, K. D. *Tetrahedron Lett.* 1997, *38*, 1527-1530.

[181] Bolm, C.; Maischak, A. *Synlett* 2001, 93-95.

[182] Wöltinger, J.; Henniges, H.; Krimmer, H.-P.; Bommarius, A. S.; Drauz, K. *Tetrahedron: Asymmetry* 2001, *12*, 2095-2098.

[183] Yang, X.-W.; Liu, H.-Q.; Xu, M.-H.; Lin, G.-Q. *Tetrahedron: Asymmetry* 2004, *15*, 1915-1918.

[184] Cheng, S. K.; Zhang, S. Y.; Wang, P. A.; Kuang, Y. Q.; Sun, X. L. *Appl. Organometal. Chem.* 2005,*19*, 975-979.

[185] Smith, J. G. *Synthesis*, 1984, 629-656.

[186] (a) Katsuki, T. *Coord. Rev.* 1995, *140*, 189-214. (b) Jacobsen, E. N.; Wu, M. H. in Jacobsen, E. N.; Pfaltz, A., Yamamoto, H (Eds.) Comprehensive Asymmetric Catalysis Springer-Verlag, Berlin 1999, 649-677 .

[187] (a) De, B. B.; Lohray, B. B.; Sivaram, S.; Dhal, P. K. Tetrahedron: Asymmetry, 1995, *6* , 2105-2108. (b) De, B. B.; Lohray, B. B.; Sivaram, S.; Dhal, P. K. *J. Polym. Sci. A: Polym. Chem.* 1997, *35*, 1809-1818.

[188] (a) Minutolo, F.; Pini, D.; Salvadori, P. *Tetrahedron Lett.* 1996, *37*, 3375-3379. (b) Minutolo, F.; Pini, D.; Petri, A.; Salvadori, P. *Tetrahedron: Asymmetry* 1996, *7*, 2293-2302.

[189] Angelino, M. D.; Laibinis, P. E. *Macromolecules*, 1998, *31*, 7581-7587.

[190] Angelino, M. D.; Laibinis, P. E. J. *Polym. Sci., A: Polym.Chem.*, 1999, *37*, 3888-3898.

[191] Canali, L.; Deleuze, H.; Sherrington, D. C. *React. Funct. Polym.* 1999, *40*, 155-168.

[192] Canali, L.; Cowan, E.; Deleuze, H.; Gibson, C.; Sherrington, D. C. *Chem. Commun.* 1998 2561-2562

[193] Canali, L.; Cowan, E.; Deleuze, H.; Gibson, C. L.; Sherrington, D. C. *J. Chem. Soc., Perkin Trans. 1* 2000, 2055-2066.

[194] Reger, T. S.; Janda, K. D. *J. Am. Chem. Soc.* 2000, *122*, 6929-6934.

[195] Song, C. E.; Roh, E. J.; Yu, B. M.; Chi, D. Y.; Kim, S. C.; Lee, K. J. *J. Chem. Commun.*, 2000, 615-616.

[196] Yao, X.; Chen, H.; Lü, W.; Pan, G.; Hu, X.; Zheng, Z. *Tetrahedron Lett.* 2000, *41,* 10267-10271.

[197] Song, Y.; Yao, X.; Chen, H.; Pan, G.; Hu, X.; Zheng, Z. *J. Chem. Soc., Perkin Trans. 1* 2002, 870-873.

[198] Kureshy, R. I.; Khan, N. H.; Abdi, S. H. R.; Singh, S.; Ahmed, I.; Jasra, R. V. *J. Mol. Catal. A* 2004, *218*, 141-146.

[199] (a) Smith, K.; Liu, C.-H. *Chem. Comm.* 2002, 886-887. (b) Smith, K.; Liu, C.-H.; El-Hiti, G. A. *Org. Biomol. Chem.* 2006, *4*, 917-927.

[200] Disalvo, D.; Dellinger, D.B.; Gohdes, J.W. *React. Funct. Polym.* 2002, *53*, 103-112.

[201] Borriello, C.; Del Litto, R.; Panunzi, A.; Ruffo, F. *Inorg. Chem. Comm.* 2005, *8*, 717-721.

[202] Nielsen, M.; Thomsen, A. H.;Jensen, T. R.; Jakobsen, H. J.; Skibsted, J.; Gothelf, K. V. *Eur. J. Org. Chem.* 2005, 342-347.

[203] Holbach, M.; Weck, M. *J. Org. Chem.* 2006, *71*, 1825-1836.

[204] Ferrand, Y. ; Daviaud, R. ; Le Maux, P.; Simonneaux, G. *Tetrahedron: Asymmetry* 2006, *17*, 952-960.

[205] Poriel, C.; Ferrand, Y.; Le Maux, P.; Rault-Berthelot, J.; Simonneaux, G. *Synthetic Metals* 2008, *158*, 796-801.

[206] Tan, R.; Yin, D.; Yu, N.; Tao, L.; Fu, Z.; Yin, D. *J. Mol. Catal. A: Chem.* 2006, *259*, 125-132.

[207] Beigi, M.; Haag, R.; Liese, A. *Adv. Synth. Catal.* 2008, *350*, 919-925.

[208] Beigi, M.; Roller, S.; Haag, R.; Liese, A. *Eur. J. Org. Chem.* 2008, 2135-2141.

[209] Hansen, K. B.; Leighton, J. L.; Jacobsen, E. N. *J. Am. Chem. Soc.*, 1996, *118*, 10924-10925.

[210] Annis, D. A.; Jacobsen, E. N. *J. Am. Chem. Soc.*, 1999, *121*, 4147-4154.

[211] Peukert, S.; Jacobsen, E. N. *Org. Lett.* 1999, *1*, 1245-1248.

[212] Song, Y.; Chen, H.; Hu, X.; Bai, C; Zheng, Z. *Tetrahedron Lett.*, 2003, *44*, 7081-7085.

[213] Kwon, M.-A.; Kim, G. J. *Catal. Today*, 2003, *87*, 145-151.

[214] Shin, C. K.; Kim, S. J.; Kim, G. J. *Tetrahedron Lett.*, 2004, *45*, 7429-7433.

[215] Zheng, X.; Jones C. W.; Weck, M. *Chem. Eur. J.* 2006, *12*, 576-583.

[216] Benaglia, M.; Cinquini, M.; Cozzi, F.; Celentano, G. *Org. Biomol. Chem.* 2004, 3401-3407

[217] Mandoli, A.; Orlandi, S.; Pini, D.; Salvadori, P. *Tetrahedron: Asymmetry* 2004, *15*, 3233-3244.

[218] Lundgren, S.; Ihre, H.; Moberg, C. *Arkivoc* 2008, *(vi)*, 73-80.

[219] Barbarini, A.; Raimondo, M.; Muratori, M.; Sartori, G.; Sartario, R. *Tetrahedron: Asymmetry* 2004, *15*, 2467-2473.

INDEX

A

AIBN, 21, 37, 42,45,65, 74, 82, 114, 121, 122

Alkaloid, 31, 113, 114, 115, 116, 118, 119, 121, 123, 124, 125, 126, 127, 128, 158

Aminoalcohol, 21, 22, 23, 26, 30, 31, 55, 58, 59, 65, 66, 67, 69, 70, 73, 75, 76, 77, 81, 82, 83, 84, 85, 86, 87, 88, 89, 90, 91, 92, 93, 95, 97, 100, 155

Aminohydroxylation, 113, 118, 120, 127, 158

Anchored, 1, 18, 28, 34, 50, 73, 86, 87, 91, 95, 97, 102, 106, 110, 128, 157

Anchoring, 35, 36, 85, 86, 87, 89, 100, 126

ArgoGel, 56

Argonaut resin 97

Asymmetric dihydroxylation (AD), 113, 114, 115, 116, 118, 119, 121, 122, 123, 124, 125, 126, 127, 158

AzaBox, 35, 36, 40, 41, 48

Azide, 104

B

Barlos resin, 73, 74, 85, 87

Beads, 22, 26, 29, 68, 71, 75, 76, 77, 133, 145

Biphasic, 17, 132, 158

Biotin, 31

Bis(oxazoline), 35, 37, 48, 54, 55, 56, 110

Box, 35, 36, 37, 40, 41, 42, 47, 50, 55, 57, 153

C

Camphor, 77, 97

Carbene, 33, 35, 36, 48, 54

Cavity, 2, 20, 137, 157

CBS catalyst (Corey-Bakshi-Shibata) catalyst, 26,29,158

Chemzyme, 127, 142

Chiral pocket, 19

Chitosan, 52

Chloramine T, 118

Cinchona, 31, 113, 114, 121, 124, 158

Click chemistry, 87

Continuous flow, 26, 68, 127, 154, 157

Copolymer, 23, 31, 32, 37, 39, 61, 70, 92, 95, 115, 120, 122, 123, 124, 125, 139

Copolymerization, 1, 4, 5, 7, 11, 18, 20, 28, 31, 36, 37, 41, 42, 43, 48, 58, 61, 67, 68, 70, 75, 89, 92, 97, 98, 101, 113, 117, 119, 122, 124, 130, 149

Copolymerize, 100

Copolymerized, 7, 14, 40, 41, 43, 67, 71, 95, 115, 118, 123

Crosslink, 131

Crosslinkage, 60, 72

Crosslinked, 5, 11, 12, 14, 20, 26, 28, 37, 66, 67, 68, 75, 76, 77, 89, 91, 97, 136, 137, 147, 148, 157

Crosslinker, 10, 13, 58, 60, 133, 157

Crosslinking, 1, 5, 7, 9, 20, 22, 24, 26, 29,, 30, 31, 37, 43, 45, 54, 60, 67, 72, 77, 81, 84, 91, 92, 98, 118, 119, 125, 137, 157

Cyclohexyldiamine, 11

Cyclopropanation, 2, 35, 36, 37, 40, 41, 43, 44, 47, 48, 50, 51, 52, 53, 54, 57, 140

D

Danishefsky's diene, 61

Dehydroquinine, 118, 127

Dendrimer, 37

Dendritic, 60, 92, 95, 152

Diels-Alder, 2, 55, 57, 59, 61

Dihydroquine, 116, 121

Dihydroquinidine, 115, 119, 121, 124, 126

Diisocyanate, 11, 20

Diisocyanide, 18

Disk, 26

Dithioisocyanate, 11, 12

Dithiourea, 11

Diurea, 11

Divinylbenzene, 13, 21, 66, 67, 81, 97

DP, 12

DPEN, 3, 4, 5, 7, 10, 14, 18, 106, 157

DVB, 7, 14, 21, 22, 31, 37, 42, 43, 45, 49, 50, 58, 67, 68, 69, 70, 73, 81, 82, 85, 86, 89, 91, 92, 95, 97, 98, 100, 102, 116, 131, 155, 157

E

EGDMA, 10, 20, 21, 22, 49, 130, 133, 137, 155

Electron beam, 28, 78

Electropolymerizable, 61

Electropolymerization, 50, 61

Electropolymerized, 50, 62, 141

Ephedrine, 23, 26, 66, 70, 74, 75, 76, 77, 79, 92, 93, 95, 97, 158

Epoxidation, 129, 130, 132, 133, 135, 136, 137, 140, 145, 158

Ethyleneglycol, 17, 20, 118, 119

F

Fiber, 28, 78

G

Gel, 26, 29, 43, 76, 133

Glucose, 138

GMA, 21, 22

Graft, 28, 40

Grafted, 40, 43, 48, 55, 81, 136

Grafting, 4, 5, 36, 41, 56, 75, 79, 81, 111, 130, 145

Grignard reagents, 65, 81

Grubb's catalyst, 139

H

Helical, 23

Hetero Diels-Alder, 61

Heterogeneous, 10, 31, 36, 43, 54, 62, 67, 75, 81, 82, 101, 102, 109, 113, 118, 122, 150, 157

Homogeneous, 1, 2, 3, 5, 10, 11, 13, 14, 20, 21, 23, 29, 31, 33, 35, 37, 40, 46, 50, 51, 55, 57, 61, 62, 65, 66, 72, 81, 85, 87, 117, 118, 120, 122, 130, 131, 134, 142, 146, 155, 157

Homopolymer, 37, 120, 121, 122, 139

Homopolymerization, 37, 74, 75, 150

HTR, 3, 10, 11, 12, 13, 14, 17, 18, 20, 21, 22, 23, 24, 34, 158

Hydride reduction, 3, 25, 26, 158

Hydrogenation, 3, 4, 5, 6, 7, 9, 10, 18, 158

Hydrophilic, 10, 34

Hydroquinone, 147, 149

Hydrosilylation, 3, 33

Hydroxyethylmethacrylate, 118, 119, 122, 123

I

Immobilization, 1, 14, 17, 36, 37, 40, 42, 45, 48, 57, 71, 85, 91, 102, 114, 125, 129, 130, 134

Immobilized, 3, 4, 5, 14, 18, 20, 33, 35, 36, 45, 46, 47, 54, 77, 78, 81, 83, 84, 85, 91, 97, 98, 102, 104, 109, 114, 117, 133, 134, 135, 157, 158

Imprinted, 1, 18, 20, 137

Imprinting, 18, 19, 20, 137, 157

IndaBox, 57

Iridium, 14

Iron, 140

Insertion, 37, 135, 142

J

Jacobsen catalyst, 130, 133, 134, 139, 142

K

Kinetic resolution, 2, 10, 145, 146, 147, 158

L

Leaching, 127, 143

Linear polymers, 75, 106, 107

M

Macroporous, 43, 130, 132

Merrifield, 14, 40, 41, 43, 50, 51, 72, 81, 82, 85, 86, 89, 91, 95, 104

Methacrylate, 20, 21, 23, 117, 118, 119, 122, 123

MIP, 81, 137

MMA, 155

Molybdenum, 110

Monolith, 26, 81

Monolithic, 43, 49, 81, 158

N

Norbornene, 139

Norephedrine, 23, 66, 70, 97

O

Oligomer, 158

Oxazaborolidine, 25, 26, 29, 30, 58, 59, 60

Oxazaborolidinone, 58, 59, 60

Oxazolidine, 101, 104, 105

Oxazoline, 35, 40, 46, 65, 101, 102, 104, 110, 158

P

PEG, 17, 18, 41, 47, 48, 57, 106, 111, 125, 127, 128

Pendent, 14, 23, 26, 29, 33, 120, 131

Phase transfer catalyst, 17, 142

Phthalazine, 113, 120, 122, 123, 124

Polyamide, 109

Polycondensation, 11

Polyethylene, 28, 78

polyethylene glycol, 37, 47, 102, 152

Polyhipe, 75, 76

Polymer-supported 2, 3, 6, 10, 14, 21, 26, 29, 30, 31, 34, 40, 65, 66, 71, 75, 77, 85, 86, 87, 90, 91, 92, 93, 95, 97, 98, 99, 100, 101, 102, 104, 105, 106, 110, 111, 114, 115, 116, 118, 120, 123, 125, 130, 133, 142, 150, 153, 154, 158

Polystyrene, 5, 14, 16, 26, 29, 30, 39, 41, 66, 67, 70, 72, 75, 77, 81, 84, 90, 91, 93, 102, 104, 130, 132, 136, 146, 157

Polythioureas, 11, 12, 13

Polyurea, 11, 13, 109

Polyvinyl alcohol, 99

Polyvinylpyrrolidone (PVP), 22

Porogen, 42, 43, 54, 137, 157

Porogeneous, 43

Porogenic, 50, 71, 81

Porphyrin, 48, 50, 140, 141

Precatalyst, 3, 5, 10, 51, 137

Prolinol, 26, 28, 29, 78, 89, 90, 158
PyBox, 35, 36, 40, 41, 42, 43, 45, 46, 102, 103, 104

Q

Quinidine, 113, 126
Quinine, 31, 113, 116, 128

R

Radical, 7, 14, 21, 31, 42, 43, 48, 58, 114, 119, 121, 122, 130, 150
Radical copolymerization, 7, 31, 42, 43, 48, 119, 122, 130
radical polymerization, 14
Recovered, 18, 48, 66, 70, 85, 91, 102, 104, 106, 107, 110, 125, 126, 149, 150, 154
Recoverability, 41
Recoverable, 101
Recovering, 116
Recovery, 1, 13, 50, 135, 136, 157
Recyclability, 43, 50, 68, 71, 78, 88, 104, 111, 158
Recyclable, 87
Recycle, 11, 14, 17, 22, 30, 43, 47, 63, 113, 138, 139, 142, 146, 148, 154
Recycled, 7, 70, 79, 95, 102, 120, 131, 136, 150, 154
Recycling, 1, 17, 30, 31, 32, 33, 40, 47, 50, 120, 143, 151, 158
Resin, 30, 37, 40, 41, 43, 49, 50, 56, 67, 72, 73, 74, 81, 82, 85, 86, 87, 88, 89, 91, 95, 97, 100, 102, 104, 110, 131, 132, 133, 134, 136, 138, 145, 146
Resorcinol, 37
Reusability, 101
Reuse, 1, 6, 14, 30, 34, 43, 45, 48, 52, 61, 68, 99, 135, 137, 158
Reused, 5, 10, 16, 18, 31, 37, 40, 43, 66, 71, 81, 91, 96, 101, 105, 106, 107, 110, 114, 115, 116, 122, 123, 125, 126, 127, 134, 135, 136, 138, 149, 150, 154
Reusing, 13

Rh, 10, 11, 14, 18, 19, 33, 51
Rhodium, 11, 14, 18, 20, 33, 50, 51
ROMP, 139
Ru, 3, 5, 6, 9, 12, 14, 17, 18, 20, 23, 34, 35, 43, 45, 46, 47, 48, 140
Ruthenium, 3, 12, 14, 15, 16, 20, 22, 36, 48

S

Salen, 55, 60, 61, 65, 105, 106, 129, 130, 131, 132, 133, 134, 135, 136, 137, 138, 139, 142, 146, 147, 149, 150, 152, 158
Schiff base, 35, 45, 54, 155
Spacer, 16, 39, 43, 45, 47, 48, 68, 70, 88, 90, 106, 116, 133, 142
Stabilizer, 22
Starch, 45
Styrene, 13, 14, 20, 36, 37, 40, 41, 42, 43, 45, 47, 48, 49, 50, 52, 53, 54, 58, 61, 67, 68, 70, 74, 78, 81, 92, 95, 97, 98, 100, 101, 115, 116, 130, 131, 132, 133, 134, 135, 136, 137, 138, 140, 142, 155
Sulfonamide, 14, 26, 30, 31, 65, 97, 98, 120
Suspension copolymerization, 58, 61, 67, 98, 124
Suzuki coupling, 79
Swellability, 67, 68, 123, 125
Swelled, 14, 67, 68, 118, 122
Swelling, 5, 9, 29, 56, 71, 78

T

TentaGel, 41, 50, 110, 134
Thermal polymerization, 37
Thiourea, 11
Threonine, 23, 58
Triazolium, 33
TsDPEN, 13, 14, 16, 17, 18

U

urea, 12

V

valine, 34, 58, 60
vanadium, 155

W

Wang resin, 40, 102, 138